The Expert Witness
and his Evidence

The Expert Witness and his Evidence

Second Edition

MICHAEL P. REYNOLDS
FCIArb, Solicitor
and
PHILIP S.D. KING
ARICS, Surveyor

OXFORD
BLACKWELL SCIENTIFIC PUBLICATIONS
LONDON EDINBURGH BOSTON
MELBOURNE PARIS BERLIN VIENNA

Copyright © Michael P. Reynolds &
    Philip S. D. King 1988, 1992

Blackwell Scientific Publications
Editorial Offices:
Osney Mead, Oxford OX2 0EL
25 John Street, London WC1N 2BL
23 Ainslie Place, Edinburgh EH3 6AJ
3 Cambridge Center, Cambridge,
    Massachusetts 02142, USA
54 University Street, Carlton
    Victoria 3053, Australia

Other Editorial Offices:
Librairie Arnette SA
2, rue Casimir-Delavigne
75006 Paris
France

Blackwell Wissenschafts-Verlag
Meinekestrasse 4
D-1000 Berlin 15
Germany

Blackwell MZV
Feldgasse 13
A-1238 Wien
Austria

First edition published 1988
Second edition published 1992

Set by DP Photosetting, Aylesbury, Bucks
Printed and bound in Great Britain
by Hartnolls Ltd, Bodmin, Cornwall

DISTRIBUTORS

Marston Book Services Ltd
PO Box 87
Oxford OX2 0DT
(*Orders:* Tel: 0865 791155
        Fax: 0865 791927
        Telex: 837515)

USA
Blackwell Scientific Publications, Inc.
3 Cambridge Center
Cambridge, MA 02142
(*Orders:* Tel: 800 759-6102
                617 225-0401)

Canada
Oxford University Press
70 Wynford Drive
Don Mills
Ontario M3C 1J9
(*Orders:* Tel: 416 441-2941)

Australia
Blackwell Scientific Publications
(Australia) Pty Ltd
54 University Street
Carlton, Victoria 3053
(*Orders:* Tel: 03 347-0300)

A catalogue record for this book is available
from the British Library.

ISBN 0–632–03389–4

Library of Congress
Cataloging in Publication Data
Reynolds, Michael P.
    The expert witness and his evidence/
Michael P. Reynolds and Philip S.D. King. –
2nd ed.
        p.   cm.
    Includes bibliographical references
and index.
    ISBN 0–632–03389–4
    1. Evidence, Expert – Great Britain.
I. King, Philip S.D.
II. Title.
KD7521.R49   1992
347.41'067 – dc20
[344.10767]   92-11007
                                          CIP

Contents

Foreword

His Honour Judge James Fox-Andrews QC

In all developed systems of law the expert plays an important role in the resolution of a wide range of disputes. Whether the dispute arises in criminal, family or civil matters the evidence of an expert is often crucial to the determination of the dispute. Indeed in many matters, unless the Tribunal, for example a valuer, is himself an expert in the field in which the dispute arises, the evidence of an expert or experts is essential.

The role accorded to the expert differs in different countries. In the adversarial system in civil disputes it is ordinarily left to the parties to instruct their own expert and to call them as witness at trial. In the fields of construction and personal injury disputes it will be rare for each party not to call at least one expert.

The experts make a major contribution to the administration of justice. This in its turn places important responsibilities and obligations on the expert towards the Tribunal as well as to the party instructing him, and these are in my experience more appreciated and accepted in this country than in some continental countries. Indeed it is possible because of a perception on the continent that the expert may not be objective, that the party's expert has little or no role to play there.

The expert in this country therefore has not only to acquire an expertise in his particular field, but he also has to acquire a degree of skill in acting as an expert – no easy skill to acquire. It is here that the second edition of this book particularly provides an invaluable service.

The scope of an expert's contribution is perceptibly widening in this country. In many continental systems the court expert has an extensive mediatory role.

In most construction disputes that come before Official Referees a direction is given that in the without prejudice meeting between experts, the expert should seek to narrow issues and/or agree facts. Further, by RSC Order 38 rule 38 enabling provisions now exist for the experts if possible to set out the matters on which they are agreed and the matters on which they are not agreed.

As the parties' legal advisers and their experts become more familiar with these provisions, it has become noticeable that in a significant number of cases a substantial measure of agreement has been achieved. It is too early at this stage to be sure, but I believe that as an alternative to other kinds of mediation, the time may come when in some cases the parties would wish a court to make an order for a meeting between experts to take place at a very early stage in the litigation with a view to reaching partial or complete settlement.

I am sure that anyone acting or likely to act as expert will find this book an ideal reference.

May 1992

Preface

The purpose of the Second Edition is to provide the reader with a practical guide for the expert witness, primarily concentrating on his duties and obligations. Like the First Edition of this book, the scope of the Second Edition extends slightly wider than the strict role of the expert witness to include guidance for those who act as consultants but who may subsequently become involved in the dispute process.

Since the First Edition there have been many interesting developments in the law and procedure affecting experts. These developments principally relate to 'without prejudice' meetings and the search for a means to make dispute resolution quicker and cheaper. Although mediation and conciliation are considered, arbitration remains the most popular contractual method of dispute resolution. Also included are new chapters on professional liability and negligence which are recognized as being particularly difficult areas for experts. Although such matters are primarily matters of law, the expert is given some guidance as to how he should consider such subjects and what his particular role may be. Although the work is set in the construction law context, nevertheless it retains its general application to all experts and updates the law generally.

The Scottish and European experiences are not ignored and the text refers to the use of court experts in those jurisdictions.

We are pleased to acknowledge comments given to us on the text by Judge John Newey QC and Judge James Fox-Andrews QC in relation to the work of the Official Referees' Court, and to His Honour Brian Clapham for his experiences of court experts in county courts.

We must especially thank Donald Keating QC for his valiant effort in reading through an early version of the draft manuscript and making perceptive comments; inspiring, we hope, a better text which will provide greater understanding of the complexities confronting experts in giving evidence.

We are grateful to the officials of the Law Society of England and Wales, the Law Society of Scotland, the Royal Institute of British Architects, the Royal Institution of Chartered Surveyors and the Chartered Institute of Arbitrators and other individuals and organisations who have given us the benefit of their comments on the manuscripts.

In particular, we would like to thank Christophe Dubois of the University of Paris and Susanne Renderknecht of Thummel Schutze & Partners in Stuttgart for assisting us in producing those sections dealing with the court expert in Germany and in France.

The opinions, however, expressed within the text are our own. The book, which is a guide, is no substitute for professional advice in this and the related areas of construction law, litigation and arbitration.

A second edition which is relatively different from, and more comprehensive than the first edition, would never have been written without the considerable forbearance not only of colleagues but also of families, in particular Michael's family, Rosemary, John and baby Francis, who have had to deal with this extraordinary obsession.

As always, our sincere thanks are due to Julia Burden for her patience and forbearance.

Michael P Reynolds
Philip S.D. King

1 The expert witness: role and duties

Definition

The Concise Oxford Dictionary defines an expert as a person having special skill or knowledge. This book is aimed particularly at two categories:

(1) those who are retained by a client to advise as consultant upon a matter which may subsequently become contentious; and
(2) those who are retained by the client through solicitors to advise upon a disputed matter, and who are subsequently instructed to give expert evidence by way of a written report and/or oral evidence.

Why experts are needed

Before considering the role of the expert and whether one may be sufficiently capable of acting in that role, a basic understanding of the need and necessity is required.

From early times our courts have acknowledged that the judges needed the assistance of particularly skilled persons to understand technically complex matters outside the law. Judge Newey QC (Official Referee) reminded experts of this in giving judgment in *Shell Pensions Trust Ltd* v. *Pell Frischmann & Partners (a firm)* (1986) where he considered the historical context of the expert's role in court from the time of the Renaissance and rebirth of science in Europe. He referred to Mr Justice Saunders admitting evidence of 'other Sciences or Faculties' (*Buckley* v. *Rice Thomas* (1554)). Judge Newey also referred to Lord Mansfield's judgment in *Foulkes* v. *Chadd* (1782) where he confirmed that the courts would recognize 'the Opinion of Scientific men upon proven facts may be given by men of Science within their own science'.

Since then the courts have come to recognize the role of the expert witness 'to furnish the judge with the necessary scientific criteria for testing the accuracy of their conclusions, so as to enable the judge or jury to form their own independant judgment by the application of these criteria to the facts proved in evidence'. *Davie* v. *Magistrates of Edinburgh* (1953).

In other words, the fundamental function of the expert is to assist the court by explaining technical/scientific matters so that the court fully understands the essential facts and matters in issue. This essential role should not be under- or over-estimated – the expert is simply a witness, although in a special category and his particular skill and expertise may be persuasive.

Where some experts may need guidance from lawyers is in the presentation skills required for putting the technical case forward which may successfully result in negotiation, settlement or trial. Lawyers are quite capable of giving, and indeed it is their duty to give some guidance to experts in that particular context but it is the expert himself who, with an appreciation of the duties and functions he must perform, can be the vital cog in the legal mechanism. However, lawyers must never coach an expert witness's evidence (see later discussion on *Whitehouse* v. *Jordan* (1981)).

It is vital to the understanding of that role that the expert appreciates the context in which he gives his evidence and appreciates the rules and procedures of the court before whom he appears. It must be clear to the expert at the outset that it is really a matter for the court or the arbitrator to decide upon the extent to which the expert can assist in terms of admissibility and competence and that, however eminent the expert or the lawyers presenting the case may be, the final arbiter is the judge.

Credibility

Although the expert may be described legally as the agent of, or the employee of, the clients or possibly the 'hired gun', whatever the legal or apt description is, he must have credibility. 'Credibility' means that the expert's evidence must be believable, it must be convincing, and the expert must be trustworthy. Unless the criteria of credibility are met, there is no point in an expert giving evidence.

Possibly the best description of credibility was given by Lord MacMillan who once said:

> 'I am not certain that any scientific man ought even to become partisan of one side. He may be the partisan of an opinion but ought never to accept a retainer to advocate a particular view merely because it is the view which is in the interests of the party who has retained him to maintain.
>
> To do so is to prostitute science and practise a fraud on the administration of justice. The true role of the expert witness is to offer the court the best assistance he can by getting at the *truth*. It is in some form of consultative capacity that the abilities of professional men are best utilised in the public service.'

Some scepticism about the role of the expert witness was expressed by Sir George Jessel MR in *Lord Abinger* v. *Ashton*. Although this would not represent current judical thinking, the words are quoted to demonstrate the natural suspicion of the public perception of experts:

> 'In matters of opinion I very much distrust expert evidence for several reasons. In the first place, although the evidence is given upon oath, in point of fact the person knows that he cannot be indicted for perjury, because it is only evidence as to a matter of opinion. So that you have not the authority of legal sanction. A dishonest man, knowing he could not be punished, might be inclined to judge in extravagant assertions on an occasion that required it. But that is not all. Expert evidence of this kind is evidence of persons who sometimes live by their business, but in all cases are remunerated for their evidence. An expert is not like an ordinary witness, who hoped

to get his expenses, but he is employed and paid in a sense of gain, being employed by the person who calls him. Now it is natural that his mind, however honest he may be, should be biased in favour of the person employing him, and accordingly we do find such bias . . . Undoubtedly there is a natural bias to do something serviceable for those who employ you and adequately remunerate you. It is very natural, and it is so effectual that we constantly see persons, instead of considering themselves witnesses, rather consider themselves the paid agents of the person who employs them.'

Sir George Jessel's opinion, which reflects public perception, described the temptations before experts as the hired agent of one side. It is not, however, all the expert's fault because the expert witness has to work within the confines of the adversarial system. It is probably due to the nature of the legal system, rather than due to the relationship between the employer and the expert, that the expert is moulded into the stand taken by his client. It is also obviously because he is instructed by one side and because he only has access to one side's point of view and one side's evidence. Despite this, the expert must rise above the temptation to be biased and must look objectively at the facts and make a balanced judgment.

The prospects and opportunities for bias may have been prevalent in Sir George Jessel's day but do not reflect current judicial thinking, because since those days the rules regarding experts' evidence have been changed by the requirement for the exchange of experts' reports prior to trial. Exchange is mutual and the chances of one expert surprising the other at trial with a novel point are reduced appreciably. Games of surprises are to be deplored and waste considerable time and expense during trial. The solicitor author would endorse to the full the considered opinion of Mr Justice Patrick Garland in *University of Warwick* v. *Sir Robert McAlpine* (1988) (see later).

Possibly the test of integrity comes to the fore during the trial. When an expert witness hedges his answer or hesitates, when he avoids directly answering the question or when he misinterprets the question, whether deliberately or not, the judge or arbitrator will get the impression that he is unsure of his ground and case, and raise questions as to whether the expert himself believes in it. The expert must be clear about the facts; he must have a thorough knowledge of them and must be able to demonstrate his command of the facts by giving an opinion which is directly related to those facts and not to misinterpretations of the facts or mere assumptions. Under examination he must be clearly able to demonstrate his rationalization of the issues and balanced judgment as well as a sense of fair play. Where experts often come unstuck is when the expert has not given a very full opinion in the report exchanged with the other side. The judges have given many warnings of this, e.g. Mr Michael Wright QC (as he then was) in *Kenning* v. *Eve Construction Ltd* (1989).

The expert's report: style and content

Ethics and professional conduct are very difficult matters to define. It is difficult to lay down hard and fast rules over conflicts that may occur between the expert and the lawyer. It is clear from judicial pronouncements that judges prefer to see the totality of the expert's evidence in the report, but lawyers may have a paramount interest in the interlocutory stage in trying to settle the matter. They see the report as a negotiating tool; judges see it as evidence upon which to base a judgment: therein lies the conflict.

Experts may complain that they have produced a thorough opinion, only to see it dissected by the lawyer. Experts feel that this undermines their professional integrity and that they are being treated like children, but quite often it is simply a misunderstanding of role. Counsel is on a learning curve; he is not the expert and must be educated by the expert for the trial. At the same time, the expert is not a lawyer and he is not skilled in the arts of the advocate. Counsel can see the traps into which the expert may fall, quite innocently, unless the report is tightened up. It is that misunderstanding of the roles, not necessarily any conflict of interest, which causes the problem.

Many experts do not find that counsel thoroughly or entirely rejects a report, but that the report simply requires some fine tuning for the purposes of presentation. That does not mean to say that counsel is rewriting the report, or that counsel is taking any responsibility for the opinions expressed therein.

In *Whitehouse* v. *Jordan* (1981) Lord Wilberforce said at page 256:

'It is necessary that expert evidence presented to the court should be, and should be seen to be, the independent product of the expert, uninfluenced as to form or content by the exigencies of litigation. To the extent that it is not, the evidence is likely to be not only incorrect but self defeating.'

In that case junior counsel was criticized for having drafted an expert report.

It is very often simply a matter of form. Sometimes it is a matter of expressing oneself in simple language that everyone can understand. Counsel is always anxious to ensure that there is no misunderstanding about what is written in the report.

Qualifications

The expert need not have become an expert by way of his business. In *R* v. *Silverlock* (1894) Lord Chief Justice Russell decided that a witness giving evidence as to proof of handwriting to establish a false pretence need not have become an expert in the way of his business or in any definite way. In that case it became necessary to prove that certain documents were in the defendant's handwriting. The prosecution led the evidence of a police superintendent who produced a letter and envelope which the defendant had written in his presence. He also produced a letter which the defendant admitted was in his handwriting. These documents were compared by the prosecution solicitor whose evidence was accepted and admitted by the court. The case was appealed but the Lord Chief Justice decided that the solicitor was an expert in handwriting and had gained considerable skill and experience during the course of his many years of private study of the subject. The question Lord Russell asked was whether the witness was *peritus* – was he skilled? Although it is a criminal case, it is cited here because the Lord Chief Justice was empowered to apply the civil standard of proof, i.e. on the balance of probabilities (section 8, Criminal Procedure Act 1865).

Formal qualifications are not necessary and are not a prerequisite to

giving expert evidence. Academic qualifications alone may be insufficient. Doubts were expressed about them in *Bristow* v. *Sequeville* (1850)
as to the admissibility of the opinion of experts in foreign law where the
foreign lawyer had not practised in the foreign jurisdiction. Opinion
derived from practical experience may be admitted (*Ajami* v. *Customs
Controller* (1954)).

A party can give expert evidence himself provided he is skilled, and
provided that such evidence is disclosed (*Shell Pensions Trust Ltd* v. *Pell
Frischmann & Partners (a firm)* (1986)).

Should you consider yourself an expert

If a person has obtained a special skill or knowledge in his trade,
profession or calling, or otherwise developed a particular *peritus* that
would meet Lord Russell's test, then he may consider acting as an
expert. No person should readily decide such a matter lightly. The
numerous cases of negligence against specialists of all disciplines are a
testimonial to the hazards of giving advice. We live in an increasingly
litigious world. The general public are becoming ever more litigious
and more ready to question decisions and actions of professional and
skilled men. Thus the person who puts himself forward as an expert
puts himself in the firing line, and may be subjected to rigorous and
intensive examination and some very hard and conscientious work.
Those who find it difficult to keep pace with their own practice
whether as architects, quantity surveyors, contractors or doctors, or
whatever specialism they pursue, should not venture into the unknown
without a clear understanding of the duties and obligations of the
expert. It is to that end that this book is devoted. If, however, the expert
decides to go ahead with a career as an expert witness along with his
practice and he is successful to the extent that his opinions are accepted
and prevail, and that he survives the rigours of the adversarial system,
then possibly he may find nothing more exhilarating or rewarding in
the whole of his professional career.

There is a tendency for experts to become professional experts and
even a tendency for the institutionalisation of experts. The expert who
does nothing but expert witness work can be a menace to himself, to his
client and the courts. He must persevere with his practice at all costs

and be an expert in his profession otherwise he will not understand nor be able to explain the 'state of the art' – the current accepted practice of his profession, or calling. It is little use an architect giving evidence about a design if he has never designed a similar building.

Sometimes the courts will ask a 'man of skill' as he is termed in Scotland or 'craftsman' as he is termed in England for a view, e.g. a stonemason. Such men by their long experience and skill are experts in their field and their opinion on questions arising from their trade may be admissible as expert evidence.

How experts are chosen

Experts are usually instructed and retained by the solicitor on behalf of the client. The expert, however, is not the representative or agent of the solicitor, he is a consultant acting on behalf of the client. It is advisable that his terms of engagement as an expert are settled by the solicitor and agreed in detail by the client. It is better that the client is responsible directly for payment of the expert's fees and not the solicitor unless the instructions to the solicitor expressly so require.

Solicitors, especially large firms, usually have their own lists of experts but some may seek further information from the Law Society or the appropriate professional body, e.g. the Royal Institute of British Architects (RIBA); Royal Institution of Chartered Surveyors (RICS); British Academy of Experts. Those bodies have lists of experts and are usually very helpful in giving a selection of names from which to choose. There are a considerable range of associations and institutions from whom lists of persons holding themselves out as experts may be selected. Sometimes solicitors may seek guidance from counsel as to experts. Counsel can assist in identifying particular experts who have previously advised counsel, or been examined by counsel.

There is nothing better to commend the expert than a good track record. A problem for the budding expert is that he is very possibly untried and untested. For him the best course is to seek appointments in the lower courts, usually in small claims matters, and build up by varying degrees and stages through the local county court and eventually to the High Court. Having graduated through the domestic legal and arbitration system, the expert may well go on to deal with international aribitration.

How can an expert become an expert witness?

There has been an increasing tendency in many professions to tighten up on continuing education. This has been caused partly by the trend previously noted of an increasingly litigious public. Persons who wish to become expert witnesses may attend seminars run by the Chartered Institute of Arbritrators and by some commercial organisations. These courses include mock trials, drafting reports, and working groups but are never any substitute for the hard experience of actually preparing and giving evidence in a case. It is wrong for any one to think that simply because they have acquired some recognition by an institution that is sufficient to enable them to consider themselves an expert witness. A potential expert should first of all be a good practitioner. He should then have substantial experience as a consultant and adviser. He might then after several years, depending upon the nature of that experience, be ready to deal with a contentious case. After the experience of court work the expert may wish to perfect his work by qualifying and practising as an arbitrator. Experts with good court experience tend to make good arbitrators and, vice versa, good arbitrators tend to make good experts because they can anticipate how the judge may deal with evidence in order to decide the issue of the case.

Some experts are those who have retired from practice. They are eligible to give evidence for a reasonable period of time after retirement provided they keep up to date with current practice. Usually, the consultant in the practice may be the person who has the time and experience to deal with such questions.

If the expert has a grasp of the fundamental principles of evidence and is well acquainted with expert's duties and obligations and has a mastery of his discipline and the facts, he will succeed in his task. He can learn the principles of evidence and his duties but he must practise his skill to achieve the required purpose. The principles expounded in this book may help him on his way.

Can an expert be an advocate?

The role of the expert witness is fundamentally to give an opinion as to technical matters on behalf of the client in as impartial and independent a fashion as is reasonably necessary for him to do. Expert witnesses were aptly and appropriately described recently as 'hired guns' in a perceptive article by Francis Goodall (architect) appearing in *Arbitration*, the journal of the Chartered Insititute of Arbitrators. It may have been an exaggeration but it seemed an apt description in the context in which the author placed it (JCI Arb [56] 3. 159).

The role of the advocate is to represent the client and to argue matters of law as necessary. The role of the expert is not concerned directly with matters of law; he does not argue matters of law, he argues matters of technical expertise. By virtue of the expert's professional discipline he is not trained as an advocate. He has not spent his professional career in the courts or before judges or arbitrators as a barrister or professional advocate and he will not have the same depth of skill or experience as such an advocate. To mix up these roles necessarily contradicts the role and function of the expert witness. As soon as the expert witness assumes the mantle of the advocate he becomes partisan and more directly involved on one side, whereas the expert witness ideally ought to assume an objective and impartial role. However it is appreciated that often in practice the expert takes on precisely the role described as that of the 'hired gun'.

Whilst there are excepted areas in which experts may act before planning and other tribunals, e.g. rent tribunals, it is by no means entirely clear that an expert acting as an advocate in the courts, or before an arbitrator, would enjoy the same immunity that is given to barristers or solicitors at common law.

In *Rondel* v. *Worsley* (1969) the immunity from suit of an advocate was addressed by the House of Lords. The House of Lords held that a barrister was immune from an action for negligence at the suit of a client in respect of his conduct and management of a case. The immunity appears to have been based upon public policy grounds rather than upon the absence of any contract between the client and counsel. Addressing the role of the advocate, it was said that the advocate:

(1) should be able to carry out his duty fearlessly and independently;
(2) that any action against the advocate would necessitate a retrial of the case;

but members of the Bar are obliged to accept any client, however difficult, who seeks their services.

In *Arenson* v. *Arenson* (1977) Lord Simon of Glaisdale dealt with the point of expert arbitrators and whether in given circumstances experts would have judicial immunity. He also dealt with the role of the advocate and his immunity from suit. His arguments in both cases were based on the demands of public policy. A barrister only enjoys immunity in respect of his forensic conduct since his duty to the court may conflict with and transcend his duty to his client. (See the fuller discussion of arbitrators' immunity later.)

The question of an expert acting as an advocate was also indirectly addressed by Mr Justice Garland in *University of Warwick* v. *Sir Robert McAlpine* (1988). He recalled Lord Wilberforce's words in *Whitehouse* v. *Jordan* (1981) and also said:

'In their closing speeches counsel felt it necessary to challenge not only the reliability but also the credibility of experts with unadorned attacks on their veracity. This simply should not happen where the court is called upon to decide complex scientifical technical issues. To a large extent this *excessively adversarial approach to expert evidence* could have been avoided if experts who had at various times expressed contrary or inconsistent views had dealt with this in their reports giving any necessary explanation. Similarly, where experts alter their views at a later stage or introduce a wholly new theory or interpretation, the new approach should be reduced to writing and furnished to the other parties at the earliest possible opportunity so that all the relevant experts can give the matter due consideration and, in a proper case, meet in order to define what is common ground and where they differ . . .'

Mr Justice Garland generally described the role and function of the expert in the circumstances and deplored the adversarial approach. It is possible to envisage that a conflict may arise where the expert acts as an advocate, and the 'excessively adversarial approach' to which Mr Justice Garland referred might arise but in a far more pronounced and demonstrative way.

Notwithstanding the remarks of Mr Justice Garland, the Courts and Legal Services Act 1990 makes provision for the extension of rights of audience by an authorized body designated by the Lord Chancellor under section 29 of the Act. In practice this may lead to certain types of experts applying for rights of audience. However, it may well be that the points made above will cause the authorized body to consider seriously the potential for conflicts of duty and it will need to be satisfied on these matters.

Duties

The duties described in this chapter are dealt with in greater detail in the subsequent chapters dealing with formulation of the issues and stages of preparation for trial.

The duties of an expert witness will vary in extent according to the time at which he is appointed and the details of his instructions. Usually an appointment is made, through a client's solicitor, well in advance of any hearing. This does not always mean that the expert is free from the pressure of time because the nature of his duties requires him to give early consideration to various aspects of any dispute.

The expert's duties can be divided into the following three distinct areas:

(1) initial advice,
(2) preparation for trial,
(3) hearing before court or arbitrator.

Initial advice

Once a dispute has arisen between parties it can be relatively easy to define the broad heads of claim and counterclaim, but the settlement of such a dispute is far more difficult and in most cases will depend on the facts of the case and the evidence adduced.

One of the first jobs, therefore, for the expert after appointment is to make an assessment of the matters in dispute and present to his client's legal advisors an appraisal of what he has found, for example, of any defects in construction or documentary evidence of problems in the administration of a contract, in order that the client knows at an early stage what the expert's evidence will be.

On the bases of such preliminary advice, it may be that the client is persuaded to alter his position, or alternatively he may be strengthened in his resolve to see the issue through to judgment.

One of the advantages of having an independent expert witness, as distinct from one previously involved in the matter in dispute, is that he should be able to make an entirely objective analysis of the problem and reflect such an approach in his comments.

Such an objective approach should result in both the strengths and the weaknesses of the case being identified. In fact the expert witness should be looking for the weaknesses in his client's case and make particular reference to them at this stage of giving initial advice. He will, of course, also identify the strengths of his client's case.

It is not uncommon for the expert's brief to include an early initial appraisal, which often takes the form of a preliminary report and, as such, will be a 'privileged document'. Such documents will not be made available to the other party.

Preparation for trial

Whether the issue is to be heard before an arbitrator or one of the Official Referees, the preparation for trial is likely to involve the expert in the following activities:

(1) co-operation with the client's legal team;
(2) preparation of a draft report dealing with the facts and expressing the expert's views on the issues;
(3) dialogue with the client's solicitor and barrister (counsel);
(4) 'discovery' of opponent's documents;
(5) extraction of relevant evidence from documentation and records;
(6) preparations of data for the Scott Schedule (see later);
(7) meeting with the expert appointed for the other side;
(8) negotiations towards any settlement;
(9) preparation of a final report for exchange with the opposing expert;

(10) research of support material for his expressed views; i.e., codes of practice, British Standards, byelaws;

(11) preparation of estimate for remedial works or assessment of loss and expense;

(12) the role of devil's advocate.

In any litigation or arbitration the expert witness is merely one member of a team appointed by the client to assist in resolving the dispute. The solicitor and the counsel appointed are other key members of that team. The lawyers have to present the argument before the court or tribunal and it is therefore their task to decide on tactics and approach.

The 12 activities outlined above fall conveniently into three groups. The first six relate to the expert's relationship with the other members of the client's team, in particular the instructing solicitor and counsel. The next three relate to duties in connection with the expert appointed for the other side, and the final group concern some of the activities necessary to present the expert's argument and come to conclusions before preparing the final report for exchange with the opposing side.

Discovery

It is of course essential that any person carrying out a service is clear as to the extent of his instructions, though these may not always be defined in detail in the brief given to an expert witness.

It will be necessary to inspect any documents such as correspondence, bills of quantities, drawings and contract documents which relate to the matters in dispute. The expert is usually given the opportunity to inspect such documents as are in the possession of his client and his advisors in the first instance. However, it is unlikely that all the documentation is available from this source.

By the legal process known as 'discovery', when litigation or arbitration is pending each side is entitled to inspect any relevant documentation that is in possession of the opposing party and his advisors. There are, however, a few such documents, known as 'privileged', that need not be made available to the other side.

Correspondence between the client and his solicitor, in anticipation of litigation, falls into this category.

Discovery is usually undertaken by the client's solicitor visiting the opposing party's solicitor, or perhaps the offices of the opposing party, to look through the documentation available, to make any notes and to arrange for any photocopies he may wish to obtain. It is very useful if the expert witness can be involved at this early stage, as apart from the maxim that 'two heads are better than one', the expert may well identify items of information in some of the documents which the solicitor could overlook. He is of course likely to be more familiar with documents such as construction drawings, bills of quantities or the quantity surveyor's dimension sheets.

Once the expert has satisfied himself as to the principles involved in the dispute and arrived at his own conclusions, there is normally an opportunity to confer with the instructing solicitor and barrister. By this time he may have been asked to produce a preliminary report, or a draft of a final report, which will form the basis of such discussion.

The meeting of experts

It is normal procedure for any meeting of experts from the opposing sides to be arranged in liaison with their legal advisors. Each expert must be free to make an independent appraisal of the matters in dispute but there is no reason why opposing experts should not inspect evidence together.

Where the dispute relates to alleged faulty workmanship or other elements of construction which need to be opened up, it is common practice for one expert to inform the others of any exposure he proposes, so that there is an opportunity for both to see what is revealed.

Where the matters in dispute relate only to such things as contractual issues, which will largely be assessed from a study of the documentation produced whilst the works were carried out, the opposing experts may not be asked to meet each other until they have had the opportunity to report to their clients and express their objective appraisal of the items in dispute.

When the meeting takes place the objectives will vary. Ideally the solicitor or barrister acting for each client will outline to the expert witness the purpose of any meeting they propose.

It is normal procedure for the discussions to be held entirely 'without prejudice'. In this atmosphere they can often be of an exploratory nature. Frequently meetings of experts are at the direction of the judge or arbitrator, following preliminary hearings before him, and in these circumstances the directions issued may specify the objective of such meetings.

In some instances the sole purpose of a meeting is to agree figures. This involves little more than an agreement that the mathematics of a claim are correct and perhaps that the rates used have been accurately transferred from a bill of quantities or other agreed reference point. Even at such meetings the expert witnesses are doing very valuable service to all parties, as it is clearly much less expensive for two people to get together to agree matters of detail, rather than leaving them in dispute until the time of a hearing when the whole team of legal and other professionals for each side will have to sit through the process while the same details are related by the witnesses before the judge or arbitrator.

Before the hearing is due it is likely that a date will have been fixed, either by agreement or by direction from the judge or arbitrator, for final reports of the expert witnesses to be exchanged. By this method each side is given advance notice simultaneously of the position which the opposing expert takes.

After final reports have been exchanged it is usually clear how wide the difference is between the parties. Frequently there are facts which are in dispute, and parties may have reached different conclusions because of a different view of the events which occurred to bring about the dispute.

As the expert witnesses acting for the separate parties should be able to take a more objective and detached view than the professionals who have been involved in the events leading up to the dispute, they are more likely to be able to narrow the gap between the parties by agreement of facts. It is not so easy to obtain a consensus of agreement where matters of principle are involved. This is largely because in many construction disputes valid arguments can be put to support alternative principles. Where agreement between the parties is not eventually reached and the dispute results in a hearing before an arbitrator or Official Referee, then the issue will be decided on the basis of which side put the most convincing argument determined by the weight of the oral and written evidence. This will lead to a judgment based on the balance of probabilities.

Preparing the final report

By the time he comes to prepare his final report, the expert witness should have had discussions with his client's solicitor and in conference with counsel. From these discussions he will have been given some guidance as to how the report is best presented. Every endeavour should be made to comply with counsel's wishes as to any particular format or order he requires.

The expert himself must be completely happy with the statements he has made and the conclusions he has drawn, and be prepared to uphold his position under cross-examination at the hearing. Because of this it is important that, if during any discussions before the preparation of the final report it is suggested that alterations are made, care must be taken not to change the sense of what is written to something which the expert is not prepared to support.

Most reports begin with the name, address, qualifications and experience of the person writing the report. This is likely to be about a page in length. Frequently this is followed by details of the expert's appointment and the basis upon which he has come to his conclusions. In this section it is always useful to list the documents, inspections, tests, drawings and other matters to which he has referred in order to come to his conclusions.

Where reference is made to authoritative literature such as codes of practice, Building Regulations, Building Research Establishment Digests, or recognized technical literature or text books, full references should be given; photocopies of extracts are often included as an appendix to the report.

Where information can be illustrated with graphs, histograms, charts or tables these can be useful as a summary of a section of the main text.

It is likely that a number of drafts will be prepared and modified prior to the final version being produced for exchange with the opposing expert.

It is good practice in early drafts to make reference to any weaknesses in the argument being put forward, together with an indication of their significance. Opinion varies as to whether this sort of data is included in the final version of the report, but there is good argument for retaining it in order to demonstrate to the judge or arbitrator that the matters in dispute have been thoroughly considered.

Such an approach may even be the deciding factor in the case as the judge or arbitrator invariably accepts the evidence of somes witnesses wholly or on some of the matters in dispute in preference to others. The fact that a balanced view is presented by one expert may sway him in that expert's favour where there is little to choose between the evidence.

When preparing the final report it is useful to make a separate note or have a folder for the purpose of collecting together copies of the supporting data that will need to be produced at the hearing to substantiate statements that have been made.

When the document is finally ready it is sent to the instructing solicitor with an additional copy for counsel. Following exchange there follows an opportunity for each expert to read his opponent's report.

Hearing before the court or arbitrator

When the points at issue are not settled out of court the matter will be brought before a judge or arbitrator in the formal hearing. The protocol of the court results in a more formal atmosphere than is often the case in an arbitration hearing. Even the layout of the chamber in which the matter is heard affects the atmosphere. Most building disputes are heard by Official Referees at St Dunstan's House, a relatively modern building with reasonable facilities.

Whatever the role the expert witness has fulfilled in preliminary discussions with his client and his advisors, when he appears in court or before an arbitrator he appears for the purpose of assisting the court. He is there solely to assist the judge or arbitrator in arriving at his judgment or award. The interests of the client who has engaged him are totally subordinate to this paramount duty.

The most important consideration is that the expert witness can speak with authority and with specialized knowledge about the particular issues in the dispute. With this principle in mind, it will be readily appreciated that it may not be the evidence of the expert witness who has the best technical qualifications which will carry the most weight. If, for example, the whole dispute revolves around the performance of the contracts manager employed by the contractor, the academic who has no experience on site may well be less persuasive and lack credibility on practical matters.

In 1970 the Law Reform Committee issued its 17th report which concentrated on the subject of 'Evidence of Opinion and Expert Evidence'. The role of the expert witness was considered in some detail and much that was recommended was incorporated in the Civil Evidence Act 1972, which suitably amended the 1968 Act. The Committee's comments on the role of an expert witness can best be summarised by quoting from paragraphs 7, 8 and 61 of their report:

'In any civil litigation the first task of the judge is to ascertain, from the material put before him by the parties, what events have happened in the past and it may be, what other events are likely to happen in the future. His second task is to form his own opinion as to whether those events are of such a character as would entitle the party complaining of them to a particular legal remedy against another party to the litigation. But this second task often involves his forming an opinion as to whether or not a person's conduct in relation to those events conformed to the standard of skill or care or candour to be expected of someone doing what that person did in the circumstances in which he did it.

If in either of these tasks it will assist the judge to be informed of the opinion of some other person on any matter upon which he has to make up his own mind, evidence of that person's opinion should, in principle, be admissible. The test should be: has the witness who expresses his opinion some relevant knowledge not shared by the judge which makes the opinion of that witness more likely to be right that the opinion of someone who does not possess that knowledge?

It frequently occurs, however, in the course of litigation that a judge has to form an opinion upon a matter which calls for some specialized knowledge or experience which he does not possess. It may be something perceptible upon physical examination, but which would be recognized only by someone possessed of special knowledge or experience, such as defects in the quality of goods or physical injuries to a human being. It may be an inference to be drawn from what has been perceived by witnesses with their physical senses, such as the cause of damage to goods or the diagnosis or prognosis of a disease or physical injury. It may be an opinion as to whether the conduct of a party conforms to a special standard of skill or care that it was his duty in the circumstances to display, as in cases of negligence in a particular profession or trade. In order that the judge may form correct opinion on matters of these kinds, it is essential that he should be provided with all relevant information about them by someone possessed of the requisite specialized skill and knowledge.

There are various ways in which information of this kind can be provided. One way is by expert assessors sitting with the judge to give him advice which is not generally disclosed (at any rate in detail) to the parties. Another would be to appoint an expert to make a report to the court and to the parties on any matter calling for specialized knowledge or experience. The third, which is the way usually adopted, is for the judge to be supplied with the relevant information by expert witnesses selected and called by the parties and subjected to the usual procedure of examination-in-chief, cross-examination and re-examination.

The function of an expert in litigation is not limited to giving evidence. He will help solicitors and counsel on his speciality, warn them of pitfalls, make suggestions as to cross-examination of witnesses, factual as well as expert. None of this, as distinct from the substance of the evidence which he himself proposes to give in chief at the trial if he is called, need be disclosed.'

In *Shell Pensions Trust Limited* v. *Pell Frischmann & Partners (a firm)* (1986) Judge Newey QC referred to the Law Reform Committee's report and emphasized the point that as a general rule no expert evidence should be admitted except as provided by the court's direction. As a result of the Law Reform Committee's report in 1970, the Rules of the Supreme Court (High Court and Court of Appeal) changed so that a party wishing to produce expert evidence at trial had to obtain a direction from the court and comply with the directions.

The expert's functions

At the time of the hearing the expert witness has three important functions which require differing skills. These are:

(1) To present his evidence by answering the questions addressed to him by counsel representing his client. The questions are likely to be based on his own report. This will be followed by cross-examination by the 'opposing' lawyer based on his own expert's views of the issues. Re-examination by his own side's lawyer may be necessary to clarify or emphasize points raised in cross-examination.

(2) To listen to the evidence presented by the other side, particularly on those matters within his expertise, and to alert the solicitor who is part of his client's team to any flaw or problems presented by the opposition's evidence.

(3) To be prepared to take part in negotiations with the opposing expert when the legal team are seeking a settlement.

Giving evidence

The evidence of an expert witness is likely to be of considerable importance at a hearing. If it were not so the client would have been advised earlier that the considerable expense of engaging an expert was unjustified. The expert witness should therefore ensure that the presentation of his evidence does justice to the importance of its content.

The following principles should be borne in mind at this stage:

(1) The responsibility for the conduct of the case rests primarily with the client's lawyer. By the time the hearing has arrived he should be fully aware of the expert's views on the issues in

dispute and of his professional opinion of the opposing evidence. It may well be that the lawyers have decided to concentrate on a particular aspect of the evidence as a tactical ploy, and the questioning in such circumstances will be framed in order to highlight that emphasis. The expert witness should avoid the temptation to enlarge his answers to the questions asked, even though he may feel frustrated that his own particular preference as to what is the most important element of his evidence is being overlooked. Ideally the expert witness should have been made aware during conference with his client's solicitor and barrister of the way the questioning is to be approached, but he cannot and must not be told what to say.

(2) Secondly he should be careful to present evidence that relates directly to the facts of the particular dispute which is the subject of the hearing. It will also be necessary to explain to the court or arbitrator the reason why any particular principle, formula or theory put forward by the other party is not applicable to the issues being heard.

(3) When responding to a question put to him by lawyers representing either party, the witness should address the judge or arbitrator when giving his answer.

(4) Speak up and speak distinctly.

(5) Answer the question posed but take care not to be side tracked by hypothetical questions unrelated to the facts of the case.

From this it will be seen that the expert needs to be clear in his thinking when preparing a report before the hearing to decide whether it is likely to be disclosed to the other side or be treated as a privileged document.

When preparing the report or an affidavit which will form the basis of his evidence, it must be remembered that the English legal system operates on what is termed an 'adversarial system'. This means that it is for the parties in the dispute to present arguments to the court which support their case, and on the basis of that evidence alone the judge or arbitrator will make his judgment or award. The expert witness must therefore present in evidence all the supporting data which will reinforce his client's case.

'Weight' of evidence

While professional qualifications may influence the weight which is attached to an expert's evidence of opinion and fact, anyone with considerable knowledge or experience of the matters in issue can appear as an expert witness.

It is not the quantity of evidence which is of prime importance but the 'weight' of such evidence that is presented to the court. The 'weight' given to the evidence presented to a judge or arbitrator will be determined by considerations such as:

(1) the witness's first hand knowledge of events;
(2) the extent of his experience in similar work;
(3) his standing in his trade or profession;
(4) the status of the book, standards or research material relied on to support the view he has taken;
(5) the judge's or arbitrator's view of the thoroughness of the witness's investigations;
(6) the credibility of the witness;
(7) the impression given of the witness's honesty and reliability.

When questions are put to the witness by the lawyer representing the opposing party they are often for the purpose of undermining the judge's confidence in that witness in respect of the above points.

The 'performance' of the expert witness under questioning from the lawyer acting for his client, and under cross-examination by the opposing lawyer could become the deciding factor in whether judgment is given for or against his client. In circumstances where there are a number of technical solutions to a problem in dispute, all of which are recognized within the industry, it will be the witness who convinces the court that the solution he offers is the more acceptable who will win the day.

Evidence of the opposing party

When listening to the evidence of witnesses called by the opposing party, the expert should pay careful attention to the technical content and principles being put forward. The lawyer who has called the witness will invariably be attempting to emphasize points which support his client's case and those which undermine the opposing expert's own evidence.

Any notes taken while the opposing witnesses are being examined should normally be passed to the solicitor acting for the expert's own client and given to counsel. Each note should be dated and marked with the time. If it is of immediate consequence, the degree of importance must be indicated. In some arbitration proceedings the informality of the hearing and the location and layout of the chamber in which the hearing takes place necessitate the passing of these notes directly to counsel. It should be remembered that the counsel's advocacy is planned and that both the content and timing of the questions posed by him are part of his tools of the trade.

Be careful not to interupt his concentration and remember that the solicitor to whom you may have passed your note will usually wait for what he judges to be the opportune time to pass it on. When a witness is likely to be under examination for some time it may well be appropriate to wait for a recess in the hearing before passing on these observations.

Negotiations for settlement

By far the majority of disputes are settled 'out of court' and do not proceed to a hearing. Of those settled many will reach agreement before coming before a judge or arbitrator and in those cases the expert is likely to have been involved with his client's team and the opposing expert in the negotiations leading to such a settlement.

Other disputes will continue to the commencement of a hearing when the lawyers' exercise of brinkmanship has not resulted in settlement earlier. Even then a large number of these cases will be determined after one or two days of hearing following further negotiations by the legal teams in the corridors of the court.

The expert witness will need to be prepared to meet his opposite number at these times, perhaps to agree and settle costings of remedial work; perhaps to arrive at a compromise specification or come to a consensus on other technical matters.

Although the title 'expert witness' conjures up the image of a person who spends most of his time in court giving evidence, in fact more than 80 per cent of his time is spent on analysis of evidence, preparation of reports and on meetings with his legal team or the opposing expert with a view to negotiating a settlement out of court. The objective is, of course, to get a satisfactory settlement and, if possible, to save the considerable costs involved in a court or arbitration hearing which consumes many man hours for which one or both the parties will eventually pay.

Once a date for the hearing is settled, the defending party is likely to consider making a payment into court. This reduces considerably the extent of his liability for costs if the subsequent judgment or award does not exceed the amount paid into court. Determining the amount to pay into court will be an important exercise. While such payment is likely to include an amount to cover interest, the basis of the figure agreed will be the estimated damage verifiable by means of the evidence, perhaps modified upwards to the forecast damages likely to be awarded by the judge or arbitrator. The expert witness will be required to express his view on such costs before the legal team make the payment.

In arbitration proceedings a 'sealed offer' is made instead of 'a payment into court'. In the case of *M F King (trading as Robinsons Garage), M F King Holdings (UK) Limited* v. *Thomas McKenna Limited and Holbeach Plant Hire Limited (1990)*, a question arose as to the tactical decision of whether or not to bring a sealed offer to the notice of the arbitrator. The whole point of a sealed offer is that the arbitrator knows an offer has been made, but does not open it until he has decided liability and quantum. It is a practice that some lawyers believe to be prejudicial insofar as the fact that an offer has been made by one side may unfairly prejudice that side in the mind of the arbitrator. In *King's* case the sealed offer was not brought to the attention of the arbitrator nor was he directed to open it after he had determined all issues save costs. The arbitrator was unaware of the existence of the sealed offer or of any need to make an award not dealing with costs.

Qualities of an expert witness

It will be clear from what has already been said that the expert witness will be involved in a wide variety of activities which will require differing qualities. Few individuals will possess them all.

The list below (in no particular order of importance) will alert the expert witness to the qualities best needed to tackle the various activities. As relatively few disputes go the whole way to judgment, it will be appreciated that the qualities most beneficial when giving evidence may not be needed very often.

From a study of the list it is hoped that the reader will be able to appraise himself, and where a lack is identified, he should recognize the weakness and be alert to ensure that it is not exploited.

Useful qualities are:

- an analytical mind
- objective judgment
- recognition of the merits in alternative approaches
- concise reporting
- an interest in researching
- patience
- tact
- the ability to negotiate
- coolness under pressure
- placidness – an ability not to be easily antagonized
- realism
- convincing credibility
- relevant experience
- thorough technical knowledge including alternative approaches.

In carrying out his preliminary duties the expert witness will rely heavily on his experience of the matter in dispute, which he will need to analyse, and about which he will need to make an objective judgment. At this stage it will be helpful to recognize the merits of alternative points of view, alternative methods of construction etc.

Before coming to a conclusion on an issue, a useful yardstick is to consider what a similar professional person might have done in the circumstances.

Early recognition of the weak points of one's arguments will be helpful.

Appointment and duties of the expert witness

In practice the time at which the expert witness is appointed varies considerably. In appropriate cases an early decision will be made by the solicitors acting for the parties in dispute that they require expert evidence, if only to obtain a more objective view of the facts in the dispute from a person who has not been involved in its causes. In cases of negligence an independent expert opinion on the facts will be essential.

On the other hand, it is not unknown for an expert witness to be appointed within a few weeks of the hearing and then he must prepare himself as best he can for his role.

When an early appointment is made, the existing files or a selection of papers from them are presented to the expert by the client's solicitor so that the expert can become acquainted with the subject matter of the dispute. These will include the plaintiff's claim, specifying the allegations if the dispute has progressed to that stage. In practice most lawyers will instruct an expert at an early stage rather than leaving it to the last minute.

An analytical approach is called for as relevant factual information must be separated from the mass of data contained in the documents provided. Which facts are relevant will usually depend upon the allegations made by the opposing party, and the counter allegations of the expert's client.

The next operation is for the expert witness to apply his knowledge of the technical or professional aspects of the issues and to exercise his judgment in coming to a view on the probable outcome of the events that have been shown to have taken place. At this stage, it should be recognized that in matters of judgment it is unlikely that two persons will come to exactly the same conclusion and it therefore follows that a differing view expressed by the opposing party or his expert may not be wrong, but rather an alternative reasonable conclusion.

2 The construction expert

General

The construction expert may be:

- an architect
- a quantity surveyor
- an engineer – structural/civil/mechanical/electrical
- a building surveyor
- a project manager
- a construction manager
- an accountant/auditor
- skilled man, such as a craftsman.

Although this book is primarily aimed at the first six categories in this list, the same principles apply to all professionals acting as experts.

Before describing in any detail the scope of the construction expert's general duties, his work may be classified under the following headings:

(1) contract claims;
(2) claims involving breaches of statutory duty (including Building Regulations);
(3) negligence claims.

The substantive law on these subjects is outside the scope of this book and the construction expert is referred to the standard text books on this subject, in particular *Keating on Building Contracts*, 5th edn, by Sir Anthony May and edited by Donald Keating QC, and *Hudson's Building Contracts*, 20th edn, by Duncan Wallace QC. A brief analysis, however, of construction law as it may directly affect the expert is included here because, unlike other experts, construction experts are very closely identified with the law of contract. The contract is of the essence in construction cases and since *Anns v. Merton London Borough Council* (1978) was overruled, the contract has assumed even greater importance. A construction expert, as a prerequisite, will therefore have a practical working knowledge of the contract and a considerable interest in the following brief outline.

Contract claims

The preparation and compilation of contract claims occupies the time of most construction experts. Experience in investigation techniques, evaluating evidence and deciding what is relevant are the key skills of the claims expert. Together with these skills the expert needs to have a general knowledge of the law relating to contract claims and in certain cases a very detailed practical knowledge of the contract. The interdependence of claims experts is often of key importance to the success of the case. What is crucial is that the expert of a particular discipline appreciates the limits of his expertise in a particular field.

Claims may be divided into claims by and on behalf of the employer or developer against contractors and professional teams, and the reverse situation where the contractor or professional claims against the employer. The schedule which follows gives a guide to the particular expertise that may be required in dealing with the specific contractual claim. It is by no means exhaustive and each case must be reviewed in the context of its particular facts. Legal advice must be sought in each case to decide what type of expertise is required to prove the case on the balance of probability.

Forms described in this schedule have been abbreviated.

(1) Institute of Civil Engineers (ICE), minor works form (1988) (MWC, 1988)
    This form applies to minor engineering works as prescribed by

the Institute of Civil Engineers. The notes of guidance to MWC 1988 state that its intended use is for small risks, where the period of work does not exceed six months generally, works are simple and straightforward by nature, the value does not exceed £10,000, the contractor does not have any design responsibility, and the design of work is usually complete before tenders are invited. Nominated sub-contractors are not employed.

(2) Joint Contracts Tribunal (JCT), minor works form
This is divided into:

(a) JCT agreement for minor building works (1980 edition);
(b) Intermediate form of Building Contract (IFC 84).

JCT minor works form (1980) (1988 revision)
This form was designed for use (Practice Note (M2 August 1981)) where minor works were to be carried out for an agreed lump sum and where the architect had been appointed on behalf of the employer. The form is used where there are no detailed re-measurements but a lump sum offer based on drawings and/or specifications and/or schedules, and it applies to contracts worth up to £50,000 at 1981 prices.

(3) JCT fixed fee form for prime cost contract for building works (March 1967)
This form of contract covers specific items of works at prime cost referred to in the drawings and specification of the work. The contractor carries out such work for the prime cost plus the fixed fee stated in the third schedule.

(4) JCT 63 private with/without quantities (local authorities with/ without quantities) JCT 80
These forms are used for major new building contracts where the value exceeds £1m. Other major works such as major repairs and alterations works are carried out under JCT 80. The JCT 80 form has provision for a bill of quantities if the type of work requires it.

(5) JCT 80 with quantities

This particular form includes reference to drawings and bills of quantity. The price is a lump sum contract with interim monthly payments unless otherwise stated.

(6) JCT without quantities

This contract has drawings and specifications included. The lump sum value is given with interim payments. Interim payments are made monthly unless otherwise stated. An excellent commentary on this form of contract is given in *Keating on Building Contracts*, 5th edn.

(7) JCT with contractor's design

This contract embodies the employer's design requirements and the contractor's proposals to effect such design in construction; the contractor being liable for such design. It is a lump sum contract with interim monthly payments unless otherwise agreed.

(8) ICE 6th Edition 1991

The standard form of civil engineering contract is now in its sixth edition and is issued jointly by the Institute of Civil Engineers, the Association of Consulting Engineers and the Federation of Civil Engineering Contractors. The employer pays for the works carried out by reference to the schedule of rates or particular works which can vary according to quantity. (See *Keating on Building Contracts*, 5th edn, commentary by Professor John Uff QC.)

(9) JCT/Works/1/3. (1989)

This is a form of building contract which is used by some government departments.

(10) NSC/4

This is a standard form of nominated sub-contract (NSC/4 and NSC/4a (1980 Edition)) which is used where the sub-contractor is nominated by the architects as agent of the employer.

(11) DOM/1

This is a standard form of domestic sub-contract for use with JCT 80 local authority/private/with quantities. It is published by the National Federation of Building Trades Employers (NFBTE), the Federated Associations of Specialists and Sub-contractors (FASS) and the Confederation of Associations of Specialist Engineering Contractors (CASEC).

(12) JCT management contract 1987

This is part of the JCT works package of contracts which together with the works contracts combines to make construction planning quicker on site and construction of less risk to the management contractor. Most of the risk falls to the individual works contractors who carry out the work. Disputes can be particularly complicated; the liability of the parties relatively uncertain. This form has to date not proved popular with contractors and the alternative construction management form is preferred.

The following schedule is illustrative of the way in which the expertise of different disciplines may be required in dealing with various contractual claims.

Experts retained by employer/developer
These can include the following types of cases and expertise required:

Type of problem	Clause in contract	Type of expertise
Claim for defective work	Clause 3 and 5 ICE minor works form	Engineer
	Clause 1.1 JCT minor works form (1980)	Architect/surveyor
	Clause 1.1. intermediate form building contract (IFC 84)	Architect/surveyor

Type of problem	Clause in contract	Type of expertise
	Clause 2 fixed fee form (1967 Edition)	Architect/surveyor
	Clause 2 JCT 80	Architect/surveyor/ quantity surveyor
	Clause 1 JCT 63	Architect/surveyor/ quantity surveyor
	Clause 2.10 and 4 IFC 84	Architect/surveyor
	Clauses 13.1, 13.2, 20.1, 22, 36 and 49 ICE 6th Edition	Engineer
	Clause 1.5.3 JCT management contract 1987	Architect/project manager/building/ quantity surveyor
Claim for liquidated damages and damages for non-completion	Clause 22 JCT 63	Architect/building/ quantity surveyor
	Clause 24 JCT 80	Quantity surveyor
	Clause 4.2.3 IFC 1984	Quantity surveyor
	Clause 18 JCT Fixed Fee (1967 edn)	Quantity surveyor
	Clause 2.3 JCT minor works form	Quantity surveyor
	Clause 47(1) and (2) ICE 6th Edition	Engineer/quantity surveyor
Claims for damages for repudiation	Clause 27.4 JCT 80	Architect/building surveyor
	Clauses 25–26 JCT 63	Quantity surveyor
	Clauses 21–22 Fixed Fee Form 1967	Architect/surveyor/ quantity surveyor
Claim for breach of statutory duty	(See below)	Surveyor/architect/ building surveyor
		Particular expert in water/gas industry etc.

Experts retained by and on behalf of contractor/sub-contractor

Type of problem	Clause in contract	Type of expertise
Claim for extension of time	Clause 23 JCT 63	Architect/building surveyor
	Clause 25 JCT 80	Architect/building surveyor
	Clause 2.3, 2.4 and 2.5 IFC 84	Architect/building surveyor
	Clause 12.6 ICE 6th Edition	Engineer
Claims for variation of contract	Clause 13.1 JCT 80	Architect/building/quantity surveyor
	Clause 3.6 IFC 84	
	Clause 11 JCT 63	Quantity surveyor
	Clause 52 ICE 6th Edition	Engineer/quantity surveyor
Claims for direct loss and/or expense	Clause 11(6), 24(1) 34(3) JCT 63	Architect/building surveyor
	Clause 26.1, 34.3 JCT 80	Quantity surveyor
	Clause 13(3) ICE 6th Edition	Quantity surveyor

Sub-contract claims (nominated)

Type of problem	Clause in contract	Type of expertise
Extension of time	Clause 11.2 NSC/4 NSC/4a	Architect/building surveyor
Direct loss and/or expense	Clause 13 NSC/4 NSC/4a	Quantity surveyor
Delayed completion by sub-contractor	Clause 12 NSC/4 NSC/4a	Architect
	Clause 59 ICE 6th Edition	Engineer

Sub-contract (domestic sub-contract)

Type of problem	Clause in contract	Type of expertise
Extension of time	Clause 11 DOM/1	Architect/building surveyor
Direct loss and/or expense	Clause 13 DOM/1	Quantity surveyor

Nominated sub-contract 1963 Edition (green form)

Type of problem	Clause in contract	Type of expertise
Main contractor and sub-contractor claims	Clause 8 Green form	Architect/building surveyor
Extension of time claims	Clause 8(b) Green form	Architect/building surveyor

Domestic sub-contract 1963 Edition (blue form)

Type of problem	Clause in contract	Type of expertise
Extension of time	Clause 9(3)	Architect/building surveyor

GC/Works 1/(3rd Edition) (1989)

Type of problem	Clause in contract	Type of expertise
Valuation of supervising officer's instructions	Clause 9 GC/Works/1	Quantity surveyor
Prolongation and disruption expenses	Clause 5.3	Quantity surveyor

Type of problem	Clause in contract	Type of expertise
Extension of time	Clause 28(2)	Architect/building surveyor
Liquidated damages	Clause 29	Architect/building surveyor

The above categories are not exhaustive and others will readily occur to the expert but it serves as a guide to the variety of disciplines involved in the claims process.

A range of expert knowledge may be required in any given case, and the expert should be aware that, although he may be instructed to prepare a loss and expense claim, he may also be required to give evidence in support of any claim for set off or on the counterclaim. These matters would, of course, be subject to report (see later).

Contractual Disputes

The above section demonstrates two types of claims:

(1) those which are claims under the contract provisions themselves; and
(2) those which arise from a breach of the contract provisions.

General Principles

Only a general knowledge of this subject is required by the construction expert. For advice on this matter the expert will be referred to the particular advice of counsel and those instructing counsel in the particular case. However, the general objective of an award of damages is to place the party whose legal rights have been violated 'in the same position, so far as money can do, as if his rights have been observed', per Lord Justice Asquith in *Victoria Laundry Limited* v. *Newman Limited* (1949).

Contract

Damages may be awarded provided that they are not considered remote. The basis for awarding damages is known as the rule in *Hadley v. Baxendale* which deals with cases both where the aggrieved party has no special knowledge and where he has special knowledge. Under this rule damages are such as may 'fairly and reasonably be considered either:

(1) arising naturally, (from the breach of contract) . . . or
(2) such as may reasonably be supposed to have been in contemplation of both parties at the time they made the contract, as the probable result of the breach of it'.

Liquidated damages

Liquidated damages are a genuine predetermined estimate of the damages contemplated by the parties at the time they enter into the contract. Where the contractor is in delay, and practical completion of the buildings has not been achieved by the contractual date for completion, the employer may levy liquidated damages at the appropriate rate specified in the appendix to the contract.

The tests for liquidated damages were set out by Lord Dunedin in the House of Lords in *Dunlop Ltd* v. *New Garage Co. Ltd* (1915). In summary these principles are as follows:

(1) What the parties to the contract intended by the use of words such as 'penalty' or 'liquidated damages'. The court must look behind what the parties say to ascertain their clear intention.
(2) The essence of a penalty is a payment of money stipulated as '*in terrorem*' of the offending party, whereas the essence of liquidated damages is a genuine covenanted pre-estimate of damage.
(3) Whether a sum stipulated is a penalty or liquidated damages is a question of construction which is to be decided upon the terms and inherent circumstances of each particular contract. This is judged at the time the parties entered into that contract and *not* at the time of breach.

(4) The court will consider a sum to be a penalty if it is extravagant and unconscionable in amount in comparison with the greatest loss which could conceivably be proved to have followed from the breach.

(5) It will be held to be a penalty if the breach consists only in not paying a sum of money and a sum stipulated is a sum greater than the sum which ought to have been paid.

(6) There is a presumption that it is a penalty when a single lump sum is made payable by way of compensation on the occurrence of one or more or all of several events some of which may occasion serious damage and others trifling damage.

(7) It is no obstacle to the sum stipulated being a genuine pre-estimate of damage that the consequences of the breach are such as to make precise pre-estimation almost an impossibility. On the contrary, that is just the situation when it is probable that pre-estimated damage was the true bargain between the parties.

Torts

Damages in tort are awarded on the criteria laid down in the *Victoria Laundry* case (1949), but they must arise as a direct result of the breach of duty and have been reasonably foreseeable. Again the damages must not be too remote.

Liability for breach of statutory duty

Cases involving breach of statutory duty are very rare in construction contract matters, although the expert may often come across situations where there is evidence of a breach of statutory duty.

In certain cases a statute may create a positive duty and then it is a question of interpretation for lawyers as to whether breach of that duty has been intended to give rise to an action in tort. If it is so intended, then the plaintiff will usually bring his action for breach of statutory duty.

The expert in construction should have a working knowledge of statutes such as the Town and Country Planning Acts, the Public Heath Acts, the Housing Acts and the Building Act 1984, all of which affect construction. In any particular case he may have to investigate and acquire knowledge of statutes of particular applications.

Measure of damages

Some guidance as to how the courts evaluate the question of quantum of damages on the basis of expert evidence will be of interest to the expert.

In *Phillips* v. *Ward* (1956) the court held that the appropriate measure of damages against a negligent surveyor was the difference between the value of the property as it was described in the surveyor's negligent report and the value as it should have been described. It has been argued many times that the decision in *Phillips* v. *Ward* could not be taken as a strict rule but may need some qualification in certain circumstances. Recent cases illustrate some of the situations that can arise and question whether the decision in *Phillips* v. *Ward* can be regarded still as the general rule.

In *Hopkins and Palmer* v. *Jack Cotton Partnership* (1989) a house was purchased in 1981 by the plaintiffs for £17,400 relying on the defendant's survey. The survey failed to draw attention to cracking in the walls. The market value of the property at the time of purchase was only £12,000. When the purchasers discovered the cracking they were advised by engineers to 'wait and see'. It was not until 1984 that they were advised to take remedial action. Repair costs amounted to £14,000 in 1985, but the purchasers were awarded £14,000 as repair costs rather than a sum calculated on the difference in value principle. The judge said that the plaintiffs' true loss was the cost of the remedial work. The judge seems to have founded his decision on the basis that the plaintiffs acted quite properly and reasonably in keeping the property rather than placing it on the open market as soon as they discovered the defect.

In *Syrett* v. *Carr and Neave* (1990) a country house was valued at £300,000. Its true market value was £245,000 taking into account a major infestation by death watch beetle and dampness. The defects were not discovered until two years after purchase and the repair costs amounted to £78,000. Judge Bowsher QC, Official Referee, awarded the repair costs. He held that the difference in value was an inappropriate measure as the purchaser had no reason to make an instant sale of the property because she did not know of the defects until two years after moving in, and by then she was so heavily involved with the property that it was reasonable for her to keep it and repair it.

In *Watts and Watts* v. *Morrow* (1991) Judge Bowsher took a wider view of what would justify failure to resell. The plaintiffs had purchased a country house only to find that it required extensive repairs costing around £34,000. The difference in value between the property as the surveyor described it, and in its actual condition in need of repair was £15,000. The judge awarded the cost of repair justifying his decision on the basis that if the plaintiffs had sought to cut their losses by reselling, they would have incurred a very considerable cost in reselling and finding new property. In addition, they might have incurred a very substantial loss on the resale in very different market conditions, and when they were selling what would have become a suspect house they would have had to devote much time to the sale. Again, he found the purchasers in a very difficult position and found that they acted entirely reasonably in deciding to repair the premises rather than resell.

In both the *Syrett* and *Watts* cases Judge Bowsher awarded damages for the inconvenience and distress suffered by the purchasers. In the *Watts* case Judge Bowsher sought to bring the distress awards within the class allowed by Lord Justice Staughton in *Hayes* v. *Dodd* (1990). Lord Justice Staughton had held that a prospective buyer of a house goes to a surveyor not just to be advised on the financial advisability of one of the most important transactions of his life, but also to receive reassurance that when he buys the house he will have 'peace of mind and freedom from distress'.

An appeal was lodged by the defendants in *Watts* v. *Morrow* and the Court of Appeal held that *Syrett* had been wrongly decided and that Judge Bowsher's conclusion on the amount of damages awarded was erroneous. The proper measure of damages was the 'diminution in value'. Lord Justice Ralph Gibson held that the discoverability of the defect was irrelevant to the question of whether the measure of damages recoverable was the cost of repair, or the diminution in value (although he did reserve his opinion on the date at which the diminution in value was to be calculated). Diminution in value was therefore the appropriate measure in *Syrett* as well as in the *Watts* case.

With regard to the damages awarded for distress and inconvenience by Judge Bowsher, their Lordships came to the view that the award of £4,000 to the plaintiffs was too high and revised the awards down to £750 to each plaintiff, stressing the point that courts will only award 'modest' sums in respect of this head of damage.

Breaches by the employer

These may be classified into two types:

(1) those which affect performance, and
(2) those which affect termination and rescission before completion.

Repudiation

Repudiation is defined in *Keating on Building Contracts*, 5th edn, as:

> 'an absolute refusal by the employer to carry out his part of the contract, whether made before the works commenced or whilst they are being carried out.'

Examples of repudiation may include the grounds specified in Clause 28 JCT 80 Private with Quantities provided they are sufficiently serious. This would depend upon the facts and the relation the breach bears to the employer's contractual obligations, e.g. where the employer fails to give possession of the site provided that the breach is sufficiently serious.

Partial completion

If the employer repudiates and only part of the work has been completed the contractor may either:

(1) sue for damages; the measure of damages being the loss of profit on the unfinished work plus the value of the completed work at the contract price; or
(2) treat the contract as repudiated and claim in quasi contract on a *quantum meruit* for a reasonable price for the work carried out.

Uneconomic working

This claim is usually made where there has been delay in completion or disturbance of the contractors in the regular progress of the work. There is no fixed method, but one approach frequently used in practice is to compare legitimate labour costs expended with those contemplated at the time of the contract. In order to do that one must review the price of the labour element in the contract bills and also the labour content of records maintained by the contractor. The difference between the bills cost and the recorded cost less any other damages for any other cause is a useful starting point to the assessment of damages for uneconomic working.

Breaches by the contractor

Where the contractor fails to complete the works the employer is entitled to recover damages, the measure of which is the difference between the contract price and the amount it would actually cost the employer to complete the contract work, substantially, as originally intended in a reasonable manner at the earliest reasonable opportunity (*Radford* v. *De Froberville* (1977)).

Direct loss and/or expense

By far the most popular claim these days by contractors against employers is the claim for loss and/or expense. Where profit margins are tight and interest rates are high, both the industry and the developers are placed under intense pressure to complete construction cheaper and quicker. Contractors learn the costs of low tender bids sometimes to their cost. Nothing fundamentally has changed in the preparation and pursuit of loss and expense claims in recent years but the pressures, chiefly commercial pressures, have undoubtedly intensified. This means that there is an even greater need to interpret the words of the contract in accordance with legal principles and to get that interpretation right. That is not a matter for the expert alone; it is a matter for lawyers and experts.

Contravention of statutory provisions

Where there is a breach of contract by the employer or the contractor failing to comply with the appropriate statutory provisions, e.g. the Health and Safety at Work Act, the extent of breach in the statutory provisions concerned must be very carefully considered, not only by health and safety experts (as an example) but also by lawyers. Generally, these are matters of law and are not matters for technical experts.

Betterment

Where the contractor is in breach of contract and the plaintiff claims damages for the repair and reinstatement of the building works, damages will not be reduced for betterment if the plaintiff had no reasonable choice; see *Harbutt's Plasticine* v. *Wayne Tank & Pump Co. Ltd* (1970). This would not be the case when the award of such damages would be absurd; see *Bacon* v. *Cooper (Metals) Ltd* (1982).

Building Regulations

Whether the Building Regulations themselves give rise to a statutory duty which creates a strict liability irrespective of proving the mental element is not entirely clear, but it seems from views expressed in the Court of Appeal by Lord Justice Waller in *Taylor Woodrow Construction (Midlands) Limited* v. *Charcon Structures Limited* (1982) that a breach of the Regulations by themselves would not give rise to an action for damages for breach of statutory duty without some proof of negligence. He seemed to follow Mr Justice Woolf in *Worlock* v. *Saws (A firm) and Rushmore Borough Council* (1983) in concluding that the breach of the Building Regulations was not an absolute or strict statutory duty but a duty which was tantamount to a duty of care.

Building bye-laws

In *Perry* v. *Tendring District Council* (1984) Judge John Newey QC, Official Referee, held that breach of building bye-laws did not give rise to liability in damages. He further held that the liability imposed upon a local authority was not a strict one.

When considering these matters Vincent Powell-Smith, co-author of *The Building Regulations – Explained and Illustrated,* came to the conclusion that until section 38 of the Building Act 1984 was brought into effect, any breach of the regulations without proof of negligence did not of itself give rise to a claim for damages. That conclusion would seem to be consistent with all the relevant authorities to date.

Law of torts – negligence – summary

The expert may be pardoned for being confused about the evolution of the law of negligence and only left in awe and wonder at the ramifications of the highly theoretical argument and academic debate that has taken place in this complex but critical area of construction law. The following is neither a precise summary of the evolution of that complexity nor is it to be relied upon as any substitute for the appropriate and particular legal advice in any given set of circumstances.

In *Dutton* v. *Bognor Regis Urban District Council* (1972) it was held that a local authority building inspector owed a duty of care to the property owner to point out defective foundations. This case was followed by *Anns* v. *Merton London Borough Council* (1978) where it was held that since a local authority inspector had inspected the foundations improperly the local authority was liable.

This case was followed by *Batty* v. *Metropolitan Property Realisations Limited* (1978) where it was held that the builder was liable in damages where it was shown that there had been actual physical damage to the building or there was a present or imminent danger to the health or safety of the occupiers.

In *Junior Books* v. *Veitchi* (1983) a specialist nominated sub-contractor was held liable in tort to the building owner for pure economic loss caused by defects in the floor which the sub-contractor had laid. Many academics and others have commented that this was the so-called 'high tide of negligence'.

The decisions of *Batty* and *Junior Books* can be sharply contrasted with the analytical approach of the Australian High Court taken in *Council of the Shire of Sutherland* v. *Heyman* (1985) where it was decided that the decision in *Anns* would not be followed and that, as far as the Australian courts were concerned, the local authority owed no duty of care to a subsequent purchaser in respect of defective foundations.

In 1986, whether because of academic argument and debate or because of commercial pressure from the construction lobby, the advance of the law of negligence and tort was halted abruptly by the Privy Council who quite sensibly decided as a matter of policy that in purely commercial transactions the courts should look strictly at the contract between the parties and that that contract should exclusively determine liability. The *Anns* principle was further attacked in *Leigh and Sullivan Ltd* v. *Aliakmon Shipping Co* (1986), and in *D & F Estates Limited* v. *Church Commissioners for England* (1989) it was held that a main contractor could not be liable in negligence for the alleged negligent work of a sub-contractor where the loss suffered was pure economic loss.

This case was followed in *Murphy* v. *Brentwood District Council* (1990) where the House of Lords decided that a local authority was not liable to a property owner for economic loss caused by its negligence in approving an inadequate foundation design. The last vestiges of the *Anns* decision were thus swept away so that very generally speaking the law of contract is decisive in commercial construction cases.

Whilst it is not sensible or realistic to speculate too far in matters of law, judging by history one must caution the reader that all these cases are determined on their particular facts and whilst some general principles may evolve, every case must be proved and cases can be distinguished. However entrenched recent decisions, it may be that the future cannot be predicted with certainty.

Limitation and Damage

Patent damage is damage that must be capable of being readily seen, inspected and diagnosed, e.g. spalling concrete, cracking, rising damp or condensation through lack of adequate ventilation. Latent damage, on the other hand, is that damage which has not developed or is not manifest but hidden from view and only sometimes detectable from scientific testing or thorough analysis. It is not defined in the Latent Damage Act 1986.

The law requires that there be a direct causal connection between defect and damage and experts sometimes find the greatest difficulty in proving latent defects. The expert must appreciate this point when addressing the question of defects and damage in his report, and generally when he advises how extensive testing and sampling should be.

The importance of damage being either actual or latent is not just a question of extent but also one of timing, i.e. when did the damage occur? Technically, it may be important in considering what redress or remedial works are necessary. Legally, it is critical as to whether an action or claim may be brought in respect of that damage. Once again, although a legal matter, the expert must address his mind to its significance and seek legal advice as necessary.

So far as claims are concerned, the limitation period for actions in tort and contract differ.

Torts

Section 2 of the Limitation Act 1980 provides that a party may commence an action in tort within six years of the date of accrual of the cause of action (right to sue). In negligence claims the cause of action accrues when the damage occurs. When the damage occurs is a question of fact in such cases; *D W Moore & Co.* v. *Ferrier* (1988).

Contract

Section 5 of the Limitation Act 1980 provides that the time for an action founded on simple contract is six years from the date on which the cause of action accrues, or in the case of an action on a specialty contract (under seal by deed), the period is 12 years, in both cases from the date of breach.

Latent Damage Act 1986

This Act was passed as a result of the House of Lords decision in *Pirelli* v. *Oscar Faber & Partners* (1983). The case concerned the alleged negligent design in construction of industrial premises and in particular a defective chimney. It was held that the plaintiff's legal right of action accrued when the damage actually occurred and not, as formerly believed, when the plaintiff knew or ought reasonably to have known of the damage to the building; in other words when the plaintiff discovered, or reasonably ought to have discovered the defect.

In order to overcome the loss of such rights the Latent Damage Act 1986 provided that in cases of tort the limitation period extended for a period of three years from the date of knowledge, i.e. from the date the plaintiff was entitled to commence proceedings. There is a cut-off period of 15 years from the breach of duty, if this expires first – known as the long-stop – or otherwise whichever is the later of six years from the date of accrual of the right to sue (occurrence) or three years from the date of knowledge (discoverability).

'Knowledge' was defined in the 1986 Act to include knowledge from facts observable by the plaintiff, or from facts ascertained by him with the aid of appropriate expert advice, which it is reasonable for him to see; i.e. knowledge available to the plaintiff and/or his experts.

Effects restricted

The author and consultant editor of *Keating on Building Contracts*, 5th edn, sensibly warned that the 1986 Act will probably have little effect in building cases because the damage in question is either damage to the building itself (as in *Pirelli*) or irrecoverable economic loss. See *Keating on Building Contracts*, 5th edn, p 369.

3 Professional liability and the expert witness

The expert's role

In general terms there are two aspects of professional liability of concern to the expert:

(1) the extent to which the expert may advise on the breach/breaches of the appropriate professional standard of care in contract and tort; and

(2) the degree and standard of care that the expert himself must observe in carrying out his duties in contract and in tort.

It is not for the expert witness to advise a client upon matters of law. Whether a particular person is in breach of his particular duty or obligation in contract or tort or by virtue of some statutory provision is a matter for lawyers and often, in difficult borderline cases, for leading counsel. Breach of duty is a professional and a legal matter and must be treated with the attention and seriousness it deserves.

Within this context, however, the expert in many cases will have to advise as to whether X or Y fell below the professional standard that he ought to have exercised. He is permitted to give his opinion as to how and why a particular professional may have been in breach of his professional duty in relation to his professional standards. The expert must be careful not to go outside his expertise or his role and he must confine himself to the effects of the case and give his opinion within the terms of his appointment or briefing. It may be helpful if the expert appreciates the context in which his opinion may be placed, and it is in this sense, and only in this sense, that the following brief résumé, by no means exhaustive, of the law is presented.

Contractual liability

Professional liability in contract is determined by the terms of the contract itself. Terms may be express or implied. It is normal to find in contracts for professional services and in warranties express terms stating that the professional will use the appropriate standard of skill and care. Contracts with professional persons and in particular consultants (that is including warranties) are usually in writing and advisably by deed. Employers may require consultants to enter into contracts by deed so as to give the employer the benefit of a right of action for 12 years from the date of execution.

Where a professional person is required to enter into a contract in order to provide professional advice, the courts do not imply any terms that that advice is correct. The courts take the general view that most of the tasks required to be undertaken by professional men must be undertaken with reasonable care. In a contract for the supply of a service where a supplier (a person who agrees to carry out a service) is acting in the course of his business, there is an implied term that the supplier will carry out the service with reasonable skill and care; section 13 of the Supply of Goods and Services Act 1982.

The implied standard of care required in contract is similar to that in negligence. The courts may, however, intervene to question the terms of the contract where for instance one of the parties attempts to prohibit the exclusion or restriction of liability for breach. The courts will intervene to apply the test of reasonableness to that exclusion clause. Section 3 of the Unfair Contract Terms Act 1977 empowers the courts to do so. The Act also offers some guidelines to assist in defining what is reasonable but ultimately the question is one for the courts. Where the parties enter into a contract which contains terms seeking to exclude or limit liability for breach, neither party can be sure at the time of entry into the agreement that it is effectively binding.

Non-contractual liability

One of the most common and complex problems an expert will have to deal with is the tort of negligence. It is a wide area and, as Lord

McMillan once said, 'The categories of negligence are never closed.'

In law, actionable negligence is breach of a duty of care which results in damage which is not too remote a consequence of the breach. The essentials of the tort of negligence are:

(1) the existence of a duty of care;
(2) the breach of a duty;
(3) damage which is not too remote a consequence of the breach.

Damage is a difficult area. It is not so straightforward as some would suggest, but as a general guide the following categories are recognised at law as within the ambit of actionable damage:

(1) damage to property;
(2) personal injury;
(3) imminent threat to health and safety;
(4) economic loss arising from breach of professional duty (see *Hedley Byrne* v. *Heller* (1964) below).

The duty of care in negligence is owed generally to neighbours. In the leading case of *Donoghue* v. *Stevenson* (1932) Lord Atkin defined 'neighbours' as:

'persons who are so closely and directly affected by my act that I ought reasonably to have them in contemplation as being so affected when I am directing my mind as to the acts or omissions of others which are called in question.'

In *Bourhill* v. *Young* (1943) Lord Wright said that Lord Atkin had 'established a general concept of reasonable foresight as the criterion of negligence'.

Advising on breach of duty by professionals

The expert may be instructed by the solicitor to consider the professional standards and whether any person/persons may have fallen below the standard expected. In considering these he will refer to professional codes of practice or conduct, technical authorities, recent papers published on the subject and standard text books, as well as considering what a reasonably competent person in his profession would be required to do. He will also consider the 'state of the art'.

The following cases are offered as examples to the reader of how the courts have considered various professions and the duty which members of such may owe to their clients. The examples are not comprehensive. Each case depends on its facts. Although guidance may be given, the expert must consult with his client's legal advisers in considering the conclusions which the expert has reached as to the standards inferred from the evidence presented.

In *Hedley Byrne* v. *Heller and Partners* (1964) (the leading case on professional misstatement) bankers were asked about the financial stability of a customer of the bank. They gave a favourable reference, albeit with a disclaimer of responsibility. The circumstance of the enquiry made it clear to the bankers that the party on whose behalf the enquiry was made wanted to know if they could safely extend credit to the bank's customer in a substantial sum. Acting on the reference given, the plaintiffs extended credit to the bank's customer who in due course defaulted. The House of Lords held that in negligence, misrepresentation, though honest – spoken or written – may give rise to an action for damages for financial loss (pure economic loss) quite apart from the existence of any contract or fiduciary relationship. The essence of their Lordships' opinion was that the law will imply a duty of care when a party is seeking information from a party possessed of a special skill and trusts him to exercise due care, and that party knew or ought to have known that *reliance* was being placed upon his skill or judgment.

Recently, the House of Lords heard the two appeals of *Smith* v. *Eric S Bush* and *Harris* v. *Wyre Forest District Council* (1990) where the plaintiffs in both cases were house purchasers who purchased in reliance on valuations of the properties made by surveyors acting for and on the instructions of mortgagees. In both cases the surveyors' fees were paid by the plaintiffs, and in both cases it turned out that the inspections and valuations had been negligently carried out. The properties were seriously defective so that the plaintiffs suffered financial loss.

In the case of Smith, the mortgagees were a building society, the surveyors who carried out the inspection on valuation were a firm employed by that building society and their report was shown to the plaintiff. In the case of Harris, the mortgagees were the local authority who employed a member of their own staff to carry out an inspection on the valuation. His report was not shown to the plaintiff but the plaintiff rightly assumed from the local authority's offer of a mortgage loan that the property had been professionally valued as worth at least the amount of the loan.

In both cases the terms agreed between the plaintiff and the mortgagee purported to exclude any liability on the part of the mortgagee or the surveyor for the accuracy of the valuation. The House of Lords held that in both cases the surveyor making the inspection and valuation owed a duty of care to the plaintiff house purchaser and that the contractual clauses purporting to exclude liability were struck down by virtue of section 3(2) and section 11(3) of the Unfair Contract Terms Act 1977.

The significance of these cases was that the defendant giving advice or information was fully aware of the nature of the transaction which the plaintiff had in contemplation. The defendant knew that the advice or information would be communicated to him directly or indirectly and knew that it was very likely that the plaintiff would rely on that advice or information in deciding whether or not to engage in the transaction he contemplated.

The question of reliance has recently been addressed in *Caparo Industries plc* v. *Dickman* (1990). In considering this difficult question, their Lordships considered two previous decisions.

The first decision was that of *Al Saudi Banque* v. *Clark Pixley* (1989). In this case Mr Justice Millett held that the auditors of a company owed no duty of care to a bank which lent money to the company (regardless of whether the bank was an existing creditor or a potential one) because no sufficient proximity of that relationship existed in either case between the auditor and the bank.

The second case was that of *Twomax Ltd* v. *Dickson, McFarlane and Robinson* (1982) where the Lord Ordinary held that auditors owed a duty of care to potential investors, who were not shareholders, by applying the test of whether the defendants knew, or reasonably should have foreseen at the time the accounts were audited, that a person might rely on those accounts for the purpose of deciding whether or not to take over the company and therefore would suffer loss if the accounts were inaccurate.

In considering the above authorities their Lordships came to the conclusion in *Caparo Industries plc* v. *Dickman* that the purpose of the auditors' statutory duty to prepare accounts of a public company was to enable the shareholders, as a body, to exercise their interest in the general management of the company's affairs, and not for the purpose of individual speculation with a view to profit. Accordingly, the auditors' relationship with individual shareholders did not give rise to any duty of care to shareholders as potential purchasers of shares in the company. Shareholders who, relying on negligently prepared accounts, purchase shares and suffer, have no claim in negligence against the auditors.

Caparo has been further considered in *James McNaughton Papers Group Ltd* v. *Hicks Anderson & Co.* (1991) where the Court of Appeal decided that the target company's accountants had not owed the bidder a duty of care in respect of draft accounts prepared for use by the target company in a takeover negotiation. However, distinguishing *Caparo* in *Morgan Crucible Co.* v. *Hill Samuel Bank Ltd* (1991) the Court of Appeal held that if during the course of a contested takeover bid, the directors and financial advisers of the target company made express representations after an identified bidder had emerged, intending that the bidder would rely upon those representations, then they owed the bidder a duty of care not to be negligent in making representations which might mislead him.

Caparo and *Smith* v. *Eric S Bush* (1989) were applied in *Al-Makib Investments (Jersey) Ltd* v. *Longcroft* (1990) where Mr Justice Mervyn Davies held that company directors did not owe a duty of care to shareholders or anyone who relied on a prospectus published for the purpose of a rights issue when purchasing shares in the company through the stock market.

In another recent case it was held that an insurance broker owed a non–contractual duty of care to a person he knew was to become an assignee of a policy and may be liable to him for economic loss; see *Punjab National Bank* v. *De Boinville* (1991).

The duty of care incumbent on an insurance intermediary is similar to that of a broker and includes the duty to ensure that the assured is made aware of any new and onerous or unusual terms which are conditions precedent to recovering under the policy; see *Harvest Trucking Ltd* v. *Davies* (1991).

The cases of *Caparo, McNaughton, Morgan Crucible, Punjab National Bank*, and *Harvest Trucking* will be of particular interest to financial advisers in the context of the development of the rule in *Hedley-Byrne* v. *Heller and Partners*, and of further concern to professional advisers in the insurance market who have other major concerns apart from their own liability.

The expert has been given some examples of how certain professional persons may owe a duty of care. The duty never varies, but the standard of care does, depending upon the particular skill of the person involved. The expert is entitled to form an opinion based on the facts and technical authorities as to whether a professional has fallen below the ordinary standard of care required. It is for the lawyers to advise as a matter of law whether there may be a case worth pursuing by way of litigation or arbitration. A decision to proceed is not to be made by the expert witness and it may not be made by the lawyers. It is a decision that the client, having received the appropriate technical and legal advice, must make himself.

The expert's professional liability

Just like any other professional person the expert may be liable in tort or contract for breach of his duty to his client.

Contractual obligations

Express obligations
These are dependent upon the express and implied terms of the expert's brief or letter of appointment or any contract for professional services. He may be liable: for breach of a duty to exercise the requisite degree of skill and expertise required in the particular case; for failing to carry out instructions and properly report to the client or solicitors; for failing adequately to carry out investigations and advise the client of any

necessary steps, such as seeking additional specialist advice, or taking remedial action as soon as reasonably practicable – subject to the client seeking appropriate legal advice. He may also be liable if he exceeds the scope of his instructions, e.g. attending experts' meetings without authority and settling the case for a figure in excess of the limits placed upon him by way of his instructions or beyond the scope of his advice to the client. The reader is referred to the judgment of Judges Fox-Andrews and Newey regarding the scope of the expert's authority and without prejudice meetings (see later).

Implied statutory obligations
Section 13 of the Supply of Goods and Services Act 1982 provides that:

> 'In a contract for the supply of a service where the supplier is acting in the course of a business, there is an implied term that the supplier would carry out the service with reasonable care and skill.'

For the expert this means, in effect, that he has a duty as the supplier of a service to carry out that service with reasonable skill and care. This is commensurate with the duty of care and skill required in the law of tort, and if the expert has not exercised such reasonable care and skill he can additionally be found liable for breach of that statutory duty.

The expert must also be wary of time factors. The Supply of Goods and Services Act 1982 also implies a term about timely performance. In section 14, the Act provides:

> '(1)   Where, under a contract for the supply of a service by a supplier acting in the course of a business, the time for the service to be carried out is not fixed by the contract, (but is) left to be fixed in a manner agreed by the contract or determined by the course of dealing between the parties, there is an implied term that the supplier will carry out the service within a reasonable time.
>
> (2)   What is a reasonable time is a question of fact.'

In other words the expert must be minded to be careful about producing advice and reports within a reasonable time if no specific time is laid down by those instructing him. If, for an extreme example, a civil engineer is required to give advice as to the safety of a structure, fails to produce a report within a matter of weeks, rather than months, and at the end of e.g. a four month period, the structure disintegrates, then the expert may be found to have been unreasonable in performing his service for the client.

Obligations in tort

The expert's duty of care in negligence was described by Mr Justice McNair in *Bolam* v. *Friern Hospital Management Committee* (1957) as being:

'The standard of the ordinary skilled man exercising and professing to have that special skill.'

This test has been further approved by the Judicial Committee of the Privy Council in *Chin Keow* v. *Government of Malaysia* (1967) and by the House of Lords in *Whitehouse* v. *Jordan* (1981) (see later).

Further elaboration of the test was given by Lord Diplock in *Saif Ali* v. *Sydney Mitchell & Co.* (1980) where he said:

'No matter what profession it may be, the common law does not impose on those who practise it any liability for damage resulting from what in the result turn out to have been errors of judgment, unless the error was such as no reasonably well informed and competent member of that profession could have made.'

The construction expert and indeed other professionals will be judged according to the relevant standard of care, i.e. the special skill or expertise that the expert (professional man) professes. The expert cannot crave judicial immunity in these matters. The expert is not a judge. He does not exercise a judicial, arbitral or quasi-arbitral function.

The position was made very clear in *Sutcliffe* v. *Thackrah* (1974) and in *Arenson* v. *Casson Beckman Rutley & Co.* (1975) that neither architects, as in *Sutcliffe* v. *Thackrah*, nor valuers, as in *Arenson* v. *Casson*, occupied the position of a quasi-arbitrator. The submissions presented to their

Lordships in the *Sutcliffe* v. *Thackrah* appeal to the House of Lords by Donald Keating QC were subsequently adopted by their Lordships and indeed confirmed by Lord Wheatley in *Arenson* v. *Casson*. (See later for a further discussion.)

In *Campbell* v. *Edwards* (1976) Lord Denning MR said:

'the expert does not act in any arbitral or quasi-arbitral capacity and consequently has no immunity from suit on that basis.'

In *Palmer* v. *Durnford* (1991) Simon Tuckey QC sitting as a deputy judge of the High Court held that an expert witness could not claim immunity from suit by his clients for his actions in the course of preparing evidence for claim or a possible claim.

There is no legal authority that an expert appointed as a mediator or conciliator acts in a judicial capacity. In the authors' view quasi-arbitral capacity is a misnomer. An expert either acts as an arbitrator and so exercises a judicial function, or he does not. If not, he may be liable in tort and in contract.

Limiting the expert's responsibility

The importance of the *Hedley Byrne* decision for experts means that they must take special care when preparing reports and statements of opinion on the liability of those whom they may criticize. This book outlines in chapters 7 and 10 guidelines for experts who prepare reports in construction cases. Those general guidelines and principles apply for experts preparing reports in other disciplines. What experts must bear in mind are the three categories where liability can arise under *Hedley Byrne* v. *Heller* which are exhaustive. These are:

(1) the duty must be special or contractual; or
(2) the duty must be fiduciary; or
(3) the duty must arise from a relationship of proximity, the breach causing financial loss which causes critical damage to the person or the property of the plaintiff.

What the expert ought to bear in mind when giving advice or

writing reports is that these principles apply equally to his own position, where he is called in as a consultant to advise on, for example, defects in a building and necessary remedial works. If the expert has any doubts about his advice it is suggested he may expressly declare that the advice is based on restricted or limited information, that his report is subject to further investigation or enquiry, or that insufficient evidence has been made available to him, or that he feels that the client should not place greater reliance on his advice until further information has been obtained or even a second opinion obtained. Any such reservations should be clearly and expressly stated qualifying the report and its status, but there can be no certainty that the qualifying words will exclude liability for negligence unless the court is satisfied that the exclusion itself was fair and reasonable in all the circumstances.

In *Smith* v. *Eric S Bush* and *Harris* v. *Wyre Forest District Council* (1989) (see above), the reader will recall that, despite the purported exclusion of liability in the respective reports as to the accuracy of the information, both exclusion clauses (disclaimers) were struck down by section 3(2) and section 11(3) of the Unfair Contract Terms Act 1977. This Act is most important in considering the extent to which disclaimers or restrictions on liability may be made.

Discipline

Where an expert is asked to comment on the standards of another professional person it is practice to retain an expert of the same discipline, e.g. an architect against an architect. See *Worboys* v. *Acme Investments* (1969).

4 How the courts evaluate expert evidence

Only those who have attained a recognised expertise and standing in their profession should be called upon to act as experts. The task is a daunting one for the expert who has no witness or court experience. He may therefore be a good technical expert in his profession, but may prove a poor witness when giving evidence in court. This can be overcome by practice, by recognising some of the pitfalls that face the expert giving evidence, and by preparing the case as thoroughly as possible.

Some of the positive attributes and characteristics of experts have already been noted. But how do such qualities assist the expert in practice in giving his evidence, and how do experts appear in the eyes of the court? It has already been said that the expert must have an analytical mind, objective judgment, recognition of the merits of the case, patience, fluency, tact, coolness under pressure and a thorough grasp of the details of the case. His aim should be to be positive whenever possible, but this should not lead him to compromise his objectivity, and he should acknowledge where the evidence is weak.

Some lessons may be learned from how certain successful experts have achieved this objective in court.

The evaluation however must be set in the context of the criteria adopted by the courts.

The golden thread – fairness and mutuality

In *Shell Pensions Trust Limited* v. *Pell Frischmann & Partners* (a firm) (1986) Judge Newey QC held that the defendant could give evidence provided

he disclosed expert evidence before trial in accordance with the rules.
The relevant and most important rule for expert witnesses is Order 38
of the Rules of the Supreme Court 1965. It is most important that the
expert understands that rule and the fundamental principles or
mutuality and fairness that run through the whole procedural concept
of giving expert evidence. In this case Judge Newey drew particular
attention to the following extracts from the rule.

> 'Except with the leave of the court or where all parties agree no
> expert evidence may be adduced at the trial . . . unless the party
> seeking to adduce the evidence – (a) has applied to the court to
> determine whether a direction should be given under Rules 37, 38 or
> 41 (whichever is appropriate) and has complied with any direction
> given on the application . . .'

> '(1) Where an application is made under Rule 36(1) . . . the court
> may, having been satisfied that it is desirable to do so, direct that
> the substance of any expert evidence which is to be adduced by
> any party be disclosed in the form of a written report or reports
> to such other parties and within such period as the court may
> specify.
>
> (2) In deciding whether to give a direction under paragraph (1) the
> court shall have regard to all circumstances and may to such
> extent as it thinks fit, treat any of the following circumstances
> affording a sufficient reason for not giving a direction:
>
> > (a) that the expert evidence is or will be based to any material
> > extent upon a version of the facts in dispute between the
> > parties; or
> >
> > (b) that the expert evidence is or will be based to any material
> > extent on facts which are neither –
> >
> > > (i) ascertainable by the expert from his own powers of
> > > observation, nor
> > >
> > > (ii) within his general professional knowledge and
> > > experience.'

Thus, the very nature of expert evidence is perceived immediately from Order 38 where the person giving such evidence must be qualified by virtue of his own general professional knowledge and experience; it must be based on material facts observed by him; and any reports produced upon which a party wishes to rely must be exchanged between respective experts and parties. These concepts of *fairness* and *mutuality* are fundamental without which expert evidence cannot as a general rule be admitted.

Thoroughness in preparing evidence

The first case that may be taken to illustrate the quality of thoroughness in expert evidence is that of *Imperial College of Science and Technology* v. *Norman and Dawbarn* (1986). The legal issue in the case was limitation; when the cause of action arose and when the damage occurred. The question was determined on the evidence of what constituted relevant and significant damage, and when did it first occur. The late Judge Smout QC applied the test of Judge William Stabb QC in *Bromley London Borough Council* v. *Rush & Tomkins Limited* (1985).

The action was brought following displacement of 80 ceramic tiles forming part of the cladding of a 12 storey block at Imperial College in Kensington, London. Proceedings were instituted against both architects and contractors in May 1978 following the discovery of damage in 1977. Action against the architects was subsequently discontinued. Although originally pleaded in the alternative of contract and negligence, only allegations of negligence for bad design and lack of supervision were pursued against the architects. They defended themselves by pleading that the damage had been caused by settlement of the structure and that the claim was statute-barred.

The displaced tiles were manufactured in Sweden. They had grooved backs, providing an excellent key. They were waterproof and frost resistant but prone to slight thermal movement. The tiles were spaced a quarter of an inch apart and grouted, but the grout was not waterproof. The mortar bed, the rendering and the sub strata were prone to thermal and moisture movement substantially greater than the thermal movement of the tiles.

During the course of the works at Imperial College detailed records were kept by the clerk of works. He was concerned that the surface of the concrete beams and mullions should be so treated as to provide an adequate key to the render and that layers of render should not be of excessive thickness, that each coat should have time to dry and should likewise provide an adequate key, and that the line of concrete should be in accordance with the drawings, so that the render should be neither unduly thick nor unduly thin and the tiler therefore have adqaute room to affix properly the tiles relative to the line of the vertical column. The clerk of works was not satisfied with the plasterers and complained on numerous occasions of excessive thickness of render.

Expert advice was obtained on site from the manufacturers of bonding agents and the Building Research Establishment was consulted. The clerk of works noticed hollowness in the tiles on numerous dates in 1961.

The College took possession of the high block in July 1962. No practical completion certificate was issued. The final certificate was issued in December 1968.

In 1976 a survey of hollow tiles was carried out and the problem was found to be extensive. Test panels received remedial treatment in 1977 but in October of that year fell off at level 8.

In November of that year Building Research Advisory Services advised the College that the reinforced concrete substrates shrank upon drying out; cement renderings shrank unevenly and the structural concrete was found too smooth for renderings and often inadequately prepared; ceremic materials underwent low, irreversible expansion with moisture in the early years of their life cycle; mixed backgrounds made complications because of differential movements of components, and solar heat produced large reversible forces in surface membranes. The result of all this weakened the bond of the tiles, but more alarming, Building Research Advisory Services reported the existence of 'reversible strains because of the lack of overall restraint in the centre and the reduced thermal capacity of the newly separated lamina'. The consultants continued to advise the College, but in 1978 an independent expert was brought in. Assisted by a colleague, this expert put forward two alternative suggested remedial schemes, the more expensive of which was adopted. Additionally he prepared an encyclopaedia of information indicating not only precise areas of hollowness detected by a meticulous coin-tapping survey in June 1983, but listing in great detail, as the hollow tiles were removed, the nature and condition of the substrates and of the render and mortar bed immediately behind the tiles.

In his judgment the late Judge Smout said that, when he made a site inspection before the trial in February 1984 before any remedial work commenced, he noted that several areas of tiles, which sounded hollow when tapped, were extremely difficult to prize off the wall, with or without mortar or render attached, even by a skilled operative with a powered automatic chisel. He also went on to consider the possibility that shrinkage and creep of the concrete tiled frame would have had a pincer effect on the rendering, forcing the rendering and anything attached to the rendering to bow out. He also considered the likelihood of subsidence of the foundations following the hot summer of 1976, which reduced the water table beneath the site. He decided that if shrinkage and creep had been major factors, then one would have expected marked signs of failure much earlier than 1977. He found that subsidence of the foundations was not proved and found that the best evidence was the specimens of tiles that had come from the buildings.

The evidence from the tiles themselves was significant. One of the tiles that fell in October 1977 had been mostly covered by mortar on the back. However, two distinct voids in the back showed where the underside of the tile had not been buttered. One void was approximately two inches by half an inch, another approximately one inch by a quarter of an inch. On another part of the back, the render attached to the mortar sloped sharply and smoothly away to show an area of approximately two inches by one inch, which could never have been in close contact with the substrata. The voids indicated a lack of adhesion when the tile was fixed, despite the fact that it remained in position for 25 years. There was a failure between the render and the substrata. The judge also noted a pronounced pattern of black lines, not only in that tile but in other tiles exhibited and photographed which indicated widespread water penetration of the panels. One of the experts, in her analysis, calculated that out of 455 separate planes of failure, 63 per cent were between render and mortar bed, 23 per cent between concrete and substrata and render, 6 per cent between tile and mortar and 4 per cent between brickwork and render. The balance was made up for the most part of instances of failure between concrete and bed where the render had been omitted.

Judge Smout attached great importance to the expert's evidence and quoted from his elaborate investigation which gave five primary causes of failure. The judge listed these as:

'1. The entry of excessive water between the tiles and the substrata. I am satisfied that that is made out by reason of the evidence as to:

(a) defective weathering strips which were ineffective against driven rain

(b) the nature of the grout which did not waterproof the joints adjacent to the tiles and into which was driven rain and across which ran water from the face of tiles

(c) the existence of voids between the tiles and the bed (the photographs of the panels after removal of the tiles revealed marking on the render that very many tiles had been poorly buttered with mortar) and

(d) the number of dirty water marks shown in the photograph.

[The expert] suggests, as is perhaps self-evident, that the excessive water penetration resulted in internal stresses being set up by reason of frost expansion and the increased heaviness of wet render. That would lead, and as [the expert] maintains, steadily to degradation of the bond betwen bed and render and between render and substrata. . . . [The expert] recalls that some of the battens were stained and that checks after removing the tiling invariably showed the substrata to be wet after prolonged spells of dry weather. His wider experience of the site had led me to accept his view that there was an almost continuous presence of water behind the tiling. That is not to say saturation. [This view was supported by other corroborative evidence.]

2. Differential in thermal and moisture movements between tiling (with its associated bedding) and the render and likewise between tiling (with associated bedding and render) and the substrata . . .

3. Inadequate key between render and mortar bed.

4. Inadequate key between concrete and render.
5. Excessive overall thickness of bed and render setting up sheer stresses or interfaces.'

The three latter causes were corroborated in the detailed schedules previously described in the schedules as to condition and in the encyclopaedic analysis that was prepared by the experts.

Finally, completing his analysis of the expert evidence, the judge said that the facts that the experts described as causative were adequately described in their report where they stated:

'We think that it is rare for a single cause to produce failure to a point where tiles become detached and believe that such failure is normally due to a combination of two or more factors.

While we have made attempts to assess and co-relate the extent of the various causes and the manner and extent of their combination on this building, these are to a large element indefinable in terms of extent of influence of any particular cause in any particular location of failure.'

Judge Smout said:

'In the result I have come to accept that the [experts'] diagnosis is the correct one and that there were five primary causes as listed.'

This case illustrates that it is the thoroughness of surveys, the gathering of factual evidence, the analysis of the best evidence and the formation of opinions on that best evidence that provide an expert with the weight of factual evidence upon which he may base a strong opinion. Thorough preparation, coupled with convincing evidence in court and the mastery of the knowledge of the facts clearly give the expert the edge over his rival and enable him to stand up to intensive cross-examination.

Expert evidence must be the product of the expert witness

Although the solicitor has a key role in co-ordinating the team that prepares the evidence for the case, as it is he who gives the expert a brief and a timetable, and generally co-ordinates his work with that of others involved in the dispute, the lawyer cannot, and must not, write the expert's evidence for him apart, of course, from preparing a proof of evidence.

Precision and careful selection of facts

In the leading medical case of *Whitehouse* v. *Jordan and Another* in the House of Lords in 1980, Lords Wilberforce and Fraser gave a warning that, whilst some degree of consultation between experts and legal advisers was entirely proper, it is necessary that expert evidence presented to the court should be the independent product of the expert, uninfluenced as to form or content by the exigencies of litigation.

The defendant was a senior hospital registrar who took charge of the plaintiff's delivery of a baby after the mother had been in labour for a considerable time. Notes made by the consultant professor in charge of the hospital maternity unit identified the pregnancy as likely to be difficult and noted that 'a forceps' delivery would have to be tried before proceeding to delivery by Caesarian section. Noting the professor's remarks, the defendant embarked on a trial of forceps delivery. He apparently pulled on the baby six times with forceps, and following no movement, he decided to abandon the procedure and proceed to a Caesarian section. The baby was delivered by Caesarian section. The plaintiff infant sustained severe brain damage due to asphyxia. Acting through his mother, an action was brought for damages claiming that the defendant surgeon had pulled 'too hard and too long' on the plaintiff's head in carrying out the trial of forceps causing brain damage. The mother gave evidence at the trial which indicated that the forceps attempt had been inconsistent with the proper practice. At the trial, the judge found that the doctor had pulled too long and too hard, causing the head to become wedged or stuck. There was no consensus of opinion amongst the medical experts at the trial as to the meaning of the term 'impacted', whether it meant that there had been excessive or unprofessional traction with the forceps.

The evidence of the medical experts made it clear, however, that the amount of force to be properly used in a trial of forceps was a matter of clinical judgment, although there should be no attempt to pull the foetus past a boney obstruction, and if the head became so stuck as to cause asphyxia, excessive force had been used. The judge had inferred from the expert's use of the term 'disimpacted' that the plaintiff's head had become so firmly wedged or stuck in the birth canal as to indicate that excessive force had been used in the trial of forceps. The judge found that the brain damage probably occurred during the trial of forceps. He concluded that in carrying out the trial of forceps, the defendant had pulled too long and too hard so the plaintiff's head had become wedged or stuck, that in so doing or in getting the head unwedged or unstuck he had caused the plaintiff's asphyxia, and that in so using the forceps he had fallen below the standard of skill expected from an ordinary competent specialist and had therefore been negligent.

The main question before the court was whether or not the surgeon had used the ordinary skill and judgment of a competent doctor in the performance of the operation.

What was unusual in the evidence was not necessarily the technical, clinical arguments of the experts, but the plaintiff's own evidence by his mother, who said that she had been pulled with such force during the final pull. The surgeon said he had eased the head slightly upwards with forceps prior to a Caesarian section delivery.

The critical point was whether the surgeon had tried to pull past a boney obstruction contrary to the best medical practice and got the head wedged or stuck or impacted. The surgeon denied this in evidence. The question for the expert was whether, in fact, the head had got stuck or wedged. Lord Wilberforce was of the view that the surgeon had not got the head wedged or stuck. One of the expert witnesses had reported that the head was disimpacted prior to speedy delivery by Caesarian section. The word 'disimpacted' involved the courts in many hours of lengthy argument. Lord Wilberforce concluded that 'impacted' meant wedged or stuck. This would prove that the surgeon pulled too hard. According to *Steadman's Medical Dictionary*, 23rd Edition, 1976, this term was defined as 'denoting a foetus that, because of its large size or narrowing of the pelvic canal has become wedged and incapable of spontaneous advance or recession'. Three of the country's leading experts could not agree on the correct meaning of this term. In Lord Wilberforce's view, the mass of medical evidence had meant that the judge at first instance had unfortunately focussed on what Lord Wilberforce termed 'an inessential question'. Argument in his mind was not about the meaning of the word, but about *what the doctor actually did*. In his view there was insufficient evidence to have led to a finding of professional negligence.

At the time, the case aroused great controversy and publicity as to whether an error of judgment could be professional negligence. The court's view was that an error of judgment was *not* negligence; the judge at first instance had found against the doctor, but both the Court of Appeal and the House of Lords found in the doctor's favour on the basis of a review of the medical evidence.

The case illustrates that no matter how eminent the professional, the key issue and the vital question is whether he departed from the standard of care. In this case the issue was whether the doctor pulled too hard and too long, injured the child by such action and in so doing failed to pay due care and attention. A great controversy was necessarily started by the conflict of eminent opinion which can sometimes concentrate on many issues that are not necessarily of the most vital importance. The lesson for experts is that they must be most careful what they say in their reports, they must be precise and as clear as possible without giving rise to any ambiguity or misinterpretation or, if one may add, protracted debate.

Interestingly, in Lord Russell of Killowen's judgment of the case he said that an error of judgment is not in itself incompatible with negligence and he supported Lord Justice Donaldson (as he then was) in the Court of Appeal in that regard.

Special care in cases of alleged negligence

In *Maynard* v. *West Midlands RHA* (1984) a consultant physician and a consultant surgeon, recognising that the most likely diagnosis of the plaintiff's illness was tuberculosis, took a view that Hodgkin's carcinoma and sarcoidosis were also possibilities. Because Hodgkin's disease was fatal unless remedial steps were taken in its early stages, they decided that, rather than await the result of a sputum test which would involve some weeks' delay, the operation of mediantinoscopy should be performed to provide them with a biopsy. Mediantinoscopy was recognised as an operation involving a risk of damage to the left laryngeal recurrent nerve, even if properly performed, and although the operation was carried out correctly, damage did in fact occur. The biopsy proved negative.

The plaintiff brought an action against the defendant health authority for damages for negligence, contending that the decision to carry out the mediantinoscopy, rather than await the result of the sputum test, had been negligent.

At the trial, the judge said that he preferred the evidence of an expert witness called for the plaintiff who had stated that the case had almost certainly been one of tuberculosis from the outset. The Court of Appeal rejected the judge's finding, and the plaintiff subsequently appealed to the House of Lords.

The House of Lords, dismissing the appeal, held that in the medical profession, as in others, there was room for differences of opinion and practice and the court's preference for one body of opinion to another was no basis for a conclusion of negligence. The problem in the *Maynard* case was that the trial judge himself made an error, and consequently the Court of Appeal could make their own findings on the issue of negligence.

Lord Scarman quoting the judgment in *Bolam* v. *Friern Hospital Management Committee* (1957) said:

'The test is the standard of the ordinary skilled man exercising and professing that special skill. If a surgeon fails to measure up to that standard in any respect ("clinical judgment" or otherwise) he has been negligent . . .'

Lord Scarman went on to say that a case which is based on an allegation that a fully considered decision of two consultants in the field of their special skill was negligent clearly presents difficulties of proof.

In *Hunter* v. *Handley* (1955) Lord President Clyde had said:

'In the realm of diagnosis and treatment there is ample scope for genuine difference of opinion and one man clearly is not negligent merely because his conclusion differs from that of other professional men . . . The true test for establishing negligence in diagnosis or treatment on the part of a doctor is whether he has been proved to be guilty of such failure as no doctor of ordinary skill would be guilty of if acting with ordinary care . . .'

With regard to the judge's preference of one opinion to the other, Lord Scarman in *Maynard* said:

'My Lords, even if before considering the reasons given by the majority of the Court of Appeal for reversing the findings of negligence, I have to say that a judge's "preference" for one body of distinguished professional opinion to another also professionally distinguished is not sufficient to establish negligence in a practitioner whose actions have received the seal of approval of those whose opinions truthfully expressed, honestly held, were not preferred. If this was the real reason for the judge's finding he erred in law, even though elsewhere in his judgment he stated the law correctly. In the realm of diagnosis and treatment, negligence is not established by preferring one respectable body of professional opinion to another.'

The test is whether the doctor exercised the ordinary skill and competence that we would reasonably expect from a doctor. This case illustrates the point that there must be clear, factual evidence to adduce negligence which can give force and weight to the expert's opinion.

The *Bolam* test has recently been applied in *Knight* v. *Home Office* (1990) where Mr Justice Pill decided that (a) the standard of care provided for a mentally ill prisoner detained in a prison hospital was not required to be as high as the standard of care provided in a psychiatric hospital outside prison, and (b) in the particular case, the prison medical staff had not been negligent in failing to keep a prisoner under continuous observation, since their decision to observe him at 15 minute intervals was a decision which ordinary skilled medical staff in their position could have made.

In another case, *Hughes* v. *Waltham Forest Health Authority* (1990), it was held that the question for the judge in a case alleging breach of duty of care on the part of the surgeons was whether the surgeons, in reaching their decision, had displayed such a lack of clinical judgment that no surgeon exercising proper care and skill could have reached the same decision. The test in *Maynard* was applied.

Importance of demeanour in court

The case of *Joyce* v. *Yeomans* (1981) illustrates the point that in the case of evidence given by experts, the trial judge who had the opportunity to observe their demeanour was in a slightly better position than the appeal court to assess the value of evidence given. Accordingly the appeal court should be slow to interfere with judges' findings on matters of opinion evidence and even less inclined to interfere with evidence of witnesses of fact.

This slightly differs from the earlier case of *Whitehouse* v. *Jordan* where their lordships were quite content to interfere with the judge's findings at first instance.

In the *Whitehouse* case, Lord Justice Waller referred to the observations of Lord Thankerton in the well-known case of *Watt (or Thomas)* v. *Thomas* (1947) where he said:

'Where a question of fact has been tried by a judge without a jury and there is no question of misdirection of himself by the judge, an appellate court which is disposed to come to a different conclusion on the printed evidence should not do so unless it is satisfied that any advantage enjoyed by the trial judge by reason of having seen and heard the witnesses could not be sufficient to explain or justify the trial judge's conclusion ... the appellate court either because the reasons given by the trial judge are not satisfactory or because it unmistakeably appears so from the evidence, may be satisfied that it has not taken proper advantage of it having seen and heard the witnesses and the matter would then become at large for the appellate court.'

Conceding opponent's points

Interestingly Lord Justice Brandon in *Whitehouse* had some words to say as to the role of the expert witness. He said:

'Sometimes an expert witness may refuse to make what a more wise witness would make, namely proper concessions to the viewpoint of the other side. Here again, this may or may not be apparent from the transcript although plain to the trial judge.'

Interesting features of this case illustrate that if the expert evidence is so convincing, then it will carry great weight, and if of sufficient weight and successful at the first instance, an apellate court may be most reluctant to interfere with it. It seems slightly contrary to the previous case of *Maynard* where the expert evidence of both sides was equally strong.

Secondly, Lord Justice Brandon hints that it is sensible for experts sometimes to concede in proper circumstances. Often one finds experts take up intractable positions from which it is impossible for them to withdraw. Unless the position is taken up on very firm and sound evidence, the expert exposes himself to risk and his client to costs.

Is a court always bound to accept the opinion of an expert witness?

In *Davie* v. *Edinburgh Magistrates* (1953), a most important case involving the evaluation of expert opinion evidence, it was held that formal corroboration was not required in the same way as for proof of an essential fact so far as expert opinion evidence was concerned, but the court was not bound to accept the conclusions of an expert witness simply because it was uncontradicted. The court also held that a court was not entitled to rely on passages in scientific literature except insofar as that literature and opinions expressed therein had been adopted by a witness and made part of that witness's evidence or had been put to him in cross-examination.

The facts of the case were as follows. During the construction of a sewer a number of dwelling houses were damaged and their owners attributed the damage to blasting operations associated with the construction. In the action (a Scottish action) the pursuer (plaintiff or claimant) adduced no scientific opinion evidence regarding the effects of explosions on adjacent buildings, whereas the defenders (defendants) adduced the evidence of three expert witnesses in that regard. One of the experts who had made researches into the effects of blasting on nearby buildings gave evidence to the effect that the explosives used could not possibly in the circumstances have caused the damage complained of. The second expert was taken as concurring in his evidence-in-chief and in cross-examination with the first expert, but he was only cross-examined in regard to his qualification to give such evidence. The third expert gave evidence different from, and materially contradictory to the first expert. In cross-examination the first expert referred to certain pages of a scientific pamphlet in support of his views and in re-examination he was referred to that pamphlet.

At first instance the court, having rejected the expert opinion evidence and awarded damages, gave its judgment on passages in the pamphlet which had not been referred to by, or put to the witness. The defenders appealed on the basis that the pursuer had not led any expert evidence to counter the evidence of their first expert and the court was bound to accept that expert evidence; secondly the conflicting evidence of their third expert should be ignored; and thirdly the court was not entitled to have regard to part of the scientific pamphlet which had not been referred to in evidence. It was contended for the pursuer that the evidence of the first expert was uncorroborated, as the second expert who had concurred with him was not qualified to express a reliable opinion.

At first instance the defenders adduced among other witnesses a civil engineer. He gave evidence to the effect on houses of underground blasting and in particular, stated that the shock of explosion lessened according to the square of the distance from the point of explosion, and that the blasting complained of would not have caused the damage complained of owing to its distance from the house.

The defenders also led the evidence of a mining engineer and an adviser on blasting who had made a special study of the effect of blasting vibrations on structures. His evidence was to the effect that the amplitude of ground movement was proportional to the square root of the charge employed and inversely proportional to the distance of the charge from the building concerned, that the power of the explosive was relatively immaterial, and that it was its weight which was material. He further gave evidence that the amplitude of the movement permissible with safety in the case of blasting in the vicinity of houses was much greater than that which could be caused by the weight of explosives used by the defenders' contractors and that a far greater weight of explosives could have been exploded at one time without damaging the pursuer's house. In cross-examination the mining engineer was asked for the names of the standard text books on the subject on which he had given evidence and he mentioned, *inter alia.* a pamphlet 'Vibrations Due to Blasting and Their Effects on Building Structures'. Later, in re-examination, passages on pages 12 and 13 of this pamphlet were put to the witness and accepted by him as consonant with his evidence. The next witness for the defenders was Mr Sheddan who was taken as concurring in examination-in-chief and in cross-examination with the mining engineer.

When the action was appealed to the Court of Session the Lord President, Lord Cooper, had to decide whether it had been sufficiently proved at first instance that the damage to the pursuer's house was attributable to the blasting conducted by the defenders' contractors between November 1948 and December 1949. In the course of giving his judgment he said:

'That the single issue is thus one of causation; and this is an issue of pure fact for determination by the court, aided by expert opinion evidence on the technical and scientific aspects of the case . . .

On the major issue of liability the evidence falls into three chapters:

(1) factual evidence;
(2) expert evidence of an architectural or engineering nature relating to the structural damage and its probable cause; and
(3) scientific opinion evidence on the theory of propagation of ground waves caused by explosives.

Chapters (1) and (2) occupy much the greater part of the proof and include the evidence of some 19 witnesses . . . If there were nothing more in the case than this, it would be quite impossible for us to overturn the judgment of the Lord Ordinary who accepted this evidence for the inference as to causation is well nigh irrestible. I understood the defenders to accept this, and it is therefore needless to examine the details of the evidence, much of which was graphic and impressive.

The only difficulty experienced by the Lord Ordinary and developed before us arose from the scientific evidence regarding explosives and their effect. This evidence was given by Mr Teichman (civil engineer) one of the technical staff of ICI with whom a fellow employee Mr Sheddan was taken as concurring. Mr Sheddan was cross-examined on his qualifications with considerable effect, and the point was taken that Mr Teichman was duly uncorroborated. I do not consider that in the case of expert opinion evidence formal corroboration is required in the same way as it is required for proof of an essential fact, however desirable it may be in some cases to be able to rely upon two or more experts rather than upon one. The value of such evidence depends upon the authority, experience and qualifications of the expert and above all upon the extent to which his evidence carries conviction, and not upon the possibility of producing a second person to echo the sentiments of the first, usually by a formal concurrence. In this instance it would have made no difference to me if Mr Sheddan had not been adduced. The true question is whether the Lord Ordinary was entitled to discard Mr Teichman's testimony and to base his judgment upon the other evidence in the case. A founding upon the fact that no counter evidence on the science of explosives and their effects was adduced for the pursuer, the defenders went so far as to maintain that we were bound to accept the conclusions of Mr Teichman. This view I must firmly reject as contrary to the principles in accordance with which expert opinion evidence is admitted. Expert witnesses, however skilled or eminent, can give no more than evidence. They cannot usurp the functions of the jury or judge sitting as a jury, any more than a technical assessor can substitute his advice for the judgment of the Court – S.S. Bogota v. S.S. Alconda (1923) . . . Their duty is to furnish the judge or jury with the necessary scientific criteria for testing the accuracy of their conclusions, so as to enable the judge or jury to form their own independent judgment by the application of these criteria to the facts proved in evidence. The scientific opinion evidence if intelligible, convincing and tested, becomes a factor (and often an important factor) for consideration along with the whole other evidence in the case, but the decision is for the judge or jury. In particular the bare *ipse dixit* of a scientist, however eminent upon the issue in controversy, will normally carry little weight, for it cannot be tested by cross-examination nor independently appraised, and the parties have invoked the decision of a judicial tribunal and not an oracular pronouncement by an expert'.

The judge went on to say:

'Passages from a published work may be adopted by a witness and made part of his evidence or they may be put to the witness in cross-examination for his comment. But, except in so far as this is done, the court cannot in my view rely upon such works for the purpose of displacing or criticising the witness's testimony.

He concluded by saying:

'Independently of this pamphlet there is in my view amply enough in the case before us to justify fully the rejection of the explosives evidence as insufficiently vouched, unconvincing, and insufficient to displace the inference arising from the remaining evidence in the case. I am accordingly for a hearing to the interlocutor reclaimed against'.

This case has been inserted as a matter of interest for Scottish experts. It also reflects the concern of judges both in England and Scotland as to the presentation and substance of the expert evidence. Corroboration is no longer required in civil proceedings in Scotland by virtue of section 1 of the Civil Evidence (Scotland) Act 1988. It remains however a good illustration of an evaluation of expert evidence and the difficulties presented.

Damages not necessarily based on cheapest solution

In the case of *London Congregational Union Incorporated* v. *Harriss and Harriss* (1983) a new church and hall designed by defendant architects was constructed under supervision in Finchley. Construction was completed in January 1970. The hall was flooded in August 1971 as a result of a local authority storm water sewer being surcharged by the heavy downpour. Substantial damage was caused to the hall. There was no damp-proof course in the building and consequently damp penetration was caused in the basement area and the hall. The hall was not used after 1978. Judge Newey described the expert evidence as given by what he termed 'admirable witnesses' although they put forward alternative remedies to the problems of the buildings which had different cost implications. They put forward different solutions to treating open areas of the north and south side of the church hall, and also basement level which had suffered floor damage. On the north side of the open basement area experts agreed that the area should be roofed in but the type and cost of remedial work was disputed. On the south side roofing over was considered inappropriate and alternative modifications to the drainage system were proposed by the experts. The judgment is of interest in respect of the cost of remedial works where the judge says:

'The third issue is as to the damages the plaintiffs should recover. The experts, at my suggestion, with the consent of the parties, held a series of meetings to see if they could not agree figures, and we reached considerable agreement. They have put forward, however, rival schemes for dealing with the surcharging problems.

In substance, the plaintiff's expert, Mr A. says that on the north side of the building there should be constructed an aluminium and glass roof which could completely cover the three open areas and the steps prevent any water from reaching them, and that the gulleys on that side should then be blocked off. On the south side, he thinks it is impracticable to provide a sufficiently large roof or foyer to cover the whole of the open areas, as it would be prohibitively expensive, and he suggests that instead, the gulleys should be sealed, there should be provided a sump into which water would drain and there should be provided a pump, or rather two pumps, a main one and a reserve one to pump the water away. Mr M. agrees that the gutters to the north of the building be shut off. He agrees also that there should be a structure placed over the three areas at the rear, so as to prevent water getting onto the steps and into the areas, but he favours a different kind of structure, which would be considerably cheaper. It would consist partly of brick-walling and then windows, set in wooden frames, and a conventional type roof with tiles on top. Mr M. says that it would be perfectly adequate and that, what is more, it would suit the style of the building better than what is proposed by Mr A. Mr A.'s response is, however, that Mr M.'s solution would leave the plaintiffs with a permanent maintenance problem, as they would have to paint the woodwork from time to time, and I suppose deal with putty around the windows, whereas if aluminium were used, the aluminium would not need to be painted.

So far as the south side of the building is concerned, Mr M. suggests that the gulleys should not be shut off and that the pipes should remain, but they should be formed within the area of a man-hole going down to a depth of two to three feet, and that within that man-hole there should be put non-return valves which would operate only on the pipes and would be readily accessible. A person could remove the manhole top and lean down or something like that, and attend to the non-return valves or ensure that they worked properly. A cheaper solution would be achieved and I think Mr M. thinks a more satisfactory one, because he does not like pumps as they are liable to break down.'

Judge Newey, in that case, took an objective approach to the damages question, based on the cost of necessary remedial works. Readers will note that it is not the cheapest method that is always accepted and this may be rejected on the basis that an alternative, more expensive method is considered more reasonable in the circumstances. Judge Newey went on to say:

'Those are the two views. I think that so far as the north is concerned, Mr M.'s solution is the solution to be preferred. What the plaintiffs are entitled to recover is the damage flowing naturally from the defendants' breaches of duty; they are not entitled to have something that is unnecessarily better than is needed, and it seems to me that what is suggested by Mr M. will be perfectly satisfactory. I am not in a position to express any firm view as to the aesthetics but I think it is not unlikely that the M. solution will be as good or better than A.'s one in that respect. So far as the north is concerned I come down in favour of the defendant.

With regard to the south, as I have said earlier, I would not envisage members of the plaintiff congregation climbing down an eight foot man-hole. I do not suppose that the elders would much appreciate the opportunity to work regularly in a three foot man-hole. Passing round the prayer books is one thing; clambering into a man-hole is quite different. I do not think, therefore, that Mr M.'s solution is a reasonable one and I think Mr A.'s solution is the right one to adopt, namely sump with pump.'

Whilst in the first instance the judge favoured what may be described as a cheaper solution, in the second instance, the basis for the calculation of damages was a more expensive measure. The cheaper solution in this instance was deemed to be somewhat impracticable.

Summary

(1) The case is decided upon the weight of the evidence. The expert's credibility and the appreciation of his evidence will be determined by the weight the judge attaches to it, which in turn may be based on the thoroughness of the expert's surveys and preparation of reports and evidence.

(2) There is no substitute for thoroughness and dedication to the task.

(3) The expert must be convincing, he must be careful of every word he writes or says, and must be prepared to stand by it.

(4) The expert's evidence must be his evidence subject to the Rules of Evidence, unless the Rules of Evidence are excluded.

(5) The value of the expert's evidence depends upon his authority, experience and qualifications and upon its conviction – is he to be believed?

(6) Suggestions for remedial measures should be reasonable, considering all the circumstances of the case.

5 Formulation of the issues

Investigation and pleadings

As soon as a dispute arises in a construction case the parties, be they employer, contractor or architect, will call upon their advisers, in-house or independent, for advice as how best to resolve the dispute. If it is a defects case the parties will want to establish who is responsible technically and liable legally. They must call in an expert who can investigate the nature of the problem and give the client a written report which will describe the problem and suggest remedial measures.

The expert will, therefore, be asked at an early stage to diagnose the problem and its cause, sometimes to express a view, and decide who is responsible. This he usually does by way of a preliminary report. That report may be disclosed to the opposing side in the hope that they will accept the expert's findings and resolve the dispute accordingly. On the other hand they may reject the findings and seek their own independent expert's report.

The issues may be formulated only in a general way at this early stage and will gradually evolve as investigations progress and the evidence builds up.

The expert must obtain whatever evidence is available to give his preliminary opinion and must visit the site to obtain direct evidence.

The expert's aim must be to collate as much best evidence as possible. In a construction case the basic documentation he will have readily available is that in his client's possession. Depending on whether that client is the employer, architect, engineer, contractor or sub-contractor, the basic documentation will be:

- original contract, tender, bills of quantities or specification, drawings and correspondence
- Building Regulations applicable and relevant statutory provisions
- architect's instructions, certificates, variation orders, site instructions
- site diaries, daywork records, weekly reports, monthly returns, clerk of works' reports
- site meeting minutes, accounts, final accounts.

Where acting for a local authority the expert may have to obtain:

- relevant committee reports and decisions
- minutes of officers' meetings
- relevant internal memoranda between departments involved
- names of instructing officers and their location.

Whether he has to or not depends upon their strict relevance and use. Other considerations may also obtain and the expert must seek instruction from the solicitor responsible for the case.

Subsequently, the expert may require further documentation that comes into existence as a result of remedial work. It may be a major repair contract in which all the above categories of documents will be applicable.

The expert will also frequently obtain evidence from:

- surveys of the building (survey sheets and reports)
- sample testing of materials (test results)
- photographs.

At the same time as he collates this evidence, the expert should be doing his research on the matters of design and construction, getting all relevant codes of practice, articles and professional journals, government and professional reports, BRE Digests and any other relevant information; for example, manufacturers' guidelines may be of indirect interest or assistance.

The expert will be assisted in the formulation of the issues by the lawyers. If the solicitor believes there is a case for the other party to answer, he will draft instructions to counsel to settle proceedings. These may be in the form of a writ endorsed with a simple claim or fuller statement of claim alleging breach of statutory duty, nelgigence or breach of contract. Alternatively, a statement of claim can be served subsequent to the writ. At the same time as issuing a writ, the solicitor may serve a notice to concur in the appointment of an arbitrator in order to prevent a possible limitation defence where the defendant may stay the court proceedings in preference to arbitration, and a risk of being time-barred for the purposes of arbitration may arise.

The writ and statement of claim and the points of claim in arbitration may be based on the technical information and conclusions of the expert's preliminary and privileged report. Frequently, having regard to the decision in *Northern Regional Health Authority* v. *Crouch (Derek) Construction Co.* (1984), it is necessary to go to arbitration. The statement of claim and points of claim must contain precise details of the claim, the causes of the problem in dispute, the location of the defects, dates when the problems arose, the names of the parties responsible and the instances where it is alleged they were in breach of their respective legal obligations and duties.

The statement of claim may refer to the conditions of the contract, specification and bills of quantity which are in issue. For example, where the contractor is alleged to be at fault, it will plead that the contractor contracted to carry out the works shown in the drawings and described in the contract bills or specification. It will then state the facts upon which the plaintiff/claimant relies to prove certain allegations against the contractor and/or architect e.g. contractor failed to carry out or complete the works. The pleading may then allege that the contractor was under an implied duty at law to carry out the works in a good and workmanlike manner and that the contractor should have used materials which were of good quality and fit for their purpose. The statement or points of claim will then give the particulars of the defect which may closely follow the description of the expert's report as these are factual/technical matters. This will be factual and concise. Pleadings state matters of fact, not law.

It may also allege that the architect was obliged to supervise the work; prepare drawings, outline specification and detail design; obtain all necessary information from sub-contractors and suppliers; advise on tenders; make periodic site visits; issue certificates; accept buildings when completed on behalf of the employer; and that impliedly he was obliged to take reasonable care in designing the works, supervising them and carrying out all his functions. The contractor and architect's respective obligations under the Defective Premises Act 1972 may also be pleaded; for example: that the contractor owed to the plaintiffs a duty under the Defective Premises Act 1972 to see that the work taken on by them was done in a workmanlike manner with proper materials and so that as regards the work the dwellings would be fit for habitation; and the architects owed the plaintiffs under the Defective Premises Act 1972 a duty to see that the work which they took on was done in a professional manner so that the dwellings would be fit for habitation.

Example of giving particulars of defects

External works

- That the defendant contractors used mortar mix that was not of sufficient consistency and did not comply with the requirements of BS . . .
- That mortar joints to brickwork were excessively recessed resulting in the wet outer skins reducing the overall insulated value of the external walls.
- That insufficient mortar was used at perpends producing dry joints at the said perpends.
- That the brickwork in general has no adequate movement joints.
- Aware that there are movement joints but that these have been incorrectly filled with material which is insufficiently compressible for use in brickwork. This means that the joint is unable to operate properly.

Services

- That ceilings to gas-heater cupboards were omitted and the flues for the said heater units built off the first floor joists were in contravention of the local building byelaws and Building Regulations.

Site Investigations

Formulating the issues can only come after careful site investigation and analysis of the causes of the defects and other problems on site. The expert must also analyse all the documentary evidence available from his clients. The Building Research Establishment Digest defines site investigation as 'the study of site conditions to determine their probable influence on the design, construction and subsequent performance of a building'.

Where the expert is acting for the contractor, it is preferable if the preliminary site investigations are carried out by the contractor working directly under the supervision of the expert witness. The expert can then direct the contractor as to what needs to be preserved for the purposes of evidence and what samples need to be taken for testing.

For any site investigation there needs to be clearly defined objectives. What is the investigation about? This will largely depend upon what complaints have been received about the building and what damage has been caused. This will give the expert an indication of what has to be investigated and examined, for example on a major council housing estate where tenants regularly complain of dampness in walls and leaking roofs. This will immediately put the expert and his team on alert to see which properties have been affected and in what way. If the complaint is damp walls, is this caused by rising or by penetrating damp? Is the damp-proof course in good order; was it put in correctly; has it been bridged or punctured in some way? If there is penetrating damp, is it because the wall is exposed to above average weathering – is it adequately protected?

All these are matters which can start the investigation. Generally, local authorities will have abundant evidence of defects in their housing

Reporting investigation procedures

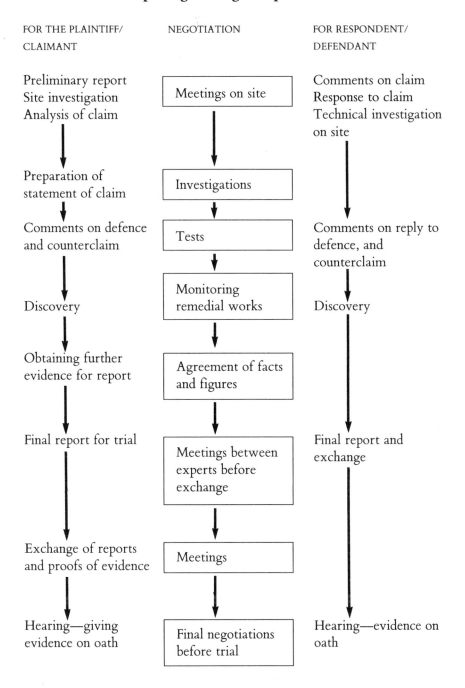

FOR THE PLAINTIFF/
CLAIMANT

NEGOTIATION

FOR RESPONDENT/
DEFENDANT

Preliminary report
Site investigation
Analysis of claim

Meetings on site

Comments on claim
Response to claim
Technical investigation
on site

Preparation of
statement of claim

Investigations

Comments on defence
and counterclaim

Tests

Comments on reply to
defence, and
counterclaim

Discovery

Monitoring
remedial works

Discovery

Obtaining further
evidence for report

Agreement of facts
and figures

Final report for trial

Meetings between
experts before
exchange

Final report and
exchange

Exchange of reports
and proofs of evidence

Meetings

Hearing—giving
evidence on oath

Final negotiations
before trial

Hearing—evidence on
oath

departments' tenants' files, in their maintenance department records or with their own architects' or engineers' departments. These are ready sources of valuable information if one is briefed on the authority's behalf.

After establishing the objectives and studying available evidence, the expert should have an initial site inspection. This may give him an idea of what is in store in future, what the costs of the investigation are likely to be, what type of contractor is required and how much opening up has to be done. The expert will be able to advise his employer on tenders for the work, the job specification and the scope of remedial works. This will usually be contained in a preliminary report or possibly a special remedial works report with approximate figures and costs for a whole scheme, or for a pilot study of various properties.

During the course of these investigations the expert may call in a specialist to examine a particular problem, for example a water services engineer to investigate any water flow problems in a council housing estate central heating system. It may be necessary to test the system in various ways, and for this specialist skills are required.

Following the site investigations, the expert will submit his findings by way of a preliminary report to his client and his solicitors, who will then advise on any necessary legal steps to be undertaken to secure redress or relief.

Summary

(1) The construction expert will examine the contract documentation and any additional documentation relating to remedial works.

(2) He will collate the evidence and research authorities and material regarding the state of the art.

(3) He will undertake all necessary preliminary and other site investigations.

(4) He will assess any damage and cost and assess requisite remedial works.

6 Procedures for resolution of disputes

Introduction

Before discussing the types of dispute resolution that are available to the construction industry in the United Kingdom, some basic points should be made on the differences between the British system, which is adversarial, and the continental system, which is inquisitiorial. A knowledge of the inquisitorial system may be of increasing importance and usefulness in this country because of the likely experimentation with aspects of inquisitorial procedures that have recently been advocated in the arbitration field.

Characteristics of the English adversarial system

The expert will usually only experience what is termed the adversarial system practised in England and Wales. He may, however, venture abroad as his experience grows and he achieves status and recognition for his particular expertise. He may also experience the development of inquisitorial techniques in arbitration matters. Should that occur, a knowledge of the main differences and characteristics of the main legal systems will be of some benefit not only to English experts giving evidence abroad, but also to those who may experience some inquisitorial innovations in the arbitration field.

The chief characteristics of the adversarial system are:

(1) Each party presents its own case and is responsible for its presentation.

(2) Each party calls its own witnesses.

(3) Each witness is subject to examination-in-chief, cross-examination and re-examination.

(4) A witness may also be subject to judicial interrogation although judges rarely intervene.

(5) The court does not call witness.

(6) The court does not enter into the arena, confining its questions to clarification of a point or seeing that fair play is carried out. Justice must be seen to be done at all times.

(7) The court hears all the evidence entered by the parties and bases its judgment on that evidence and that alone.

Characteristics of the continental civil law system

The English adversarial system is dissimilar from the continental legal system because on the continent the legal system is based largely upon Roman civil law. An awareness of the characteristics of the European system is useful, bearing in mind not only the benefits of certain inquisitorial innovations and techniques, but also the continuing influence of European Community (EC) trade custom and practice that may directly or indirectly affect trade disputes between this country and its EC partners.

The chief characteristics of the continental system are:

(1) The court will study the pleadings and all relevant evidence before the hearing.

(2) The court will decide what witnesses should be called and what experts should be appointed.

(3) The court and not counsel will call and examine witnesses. The great disadvantage under this system is that the parties have no right to call in expert evidence against the court's expert. However, an advantage is that a case is complete so far as presentation is concerned when the evidence is placed before the court.

Under the English system some argue that much time is wasted by counsel reading every document in his client's case. The authors of the *Handbook on Arbitration Practice* (1987) have argued that it would be advantageous and of considerable benefit to the client and parties if arbitrators could dispense with counsel reading every word of evidence in opening his client's case before calling any witnesses of fact. It would indeed be preferable if the arbitrator or the court could find time to read all the pleadings and all the evidence prior to the commencement of the case.

Proceedings

Under the legal system in the United Kingdom the expert witness can find himself engaged in giving evidence either in litigation or in arbitration proceedings.

So far as arbitration is concerned the forum may be composed of a single arbitrator or possibly two or three arbitrators.

In civil litigation the expert may find himself giving evidence in the county court or the High Court. He will not be concerned with any appeal matters as the appellate courts, i.e. the Court of Appeal (civil division) and the House of Lords, only have jurisdiction to review decisions of the lower courts and do not call witnesses, whether of fact or opinion.

Resolution of disputes under contract

Before a case gets as far as arbitration, there must be a dispute between the parties. Disputes may take the form of conflicts between the employer and the architect, the architect and quantity surveyor, the engineer and client and/or contractor etc. and the expert witness can find himself in any of these situations. In the writers' experience the most difficult situation can occur when the expert is faced with a conflict of interest between himself and his client, as well as being in the middle of a legitimate conflict between that client and an opposing party. This is the major reason why in arbitration experienced arbitrators with a background knowledge or experience of the private or public sectors can be of a distinct advantage in such disputes.

Arbitration

The distinct advantage of arbitration over litigation is that initially the parties are free to choose and agree their own arbitrator. They do not have this freedom of choice under the judicial system. If, however, the parties cannot agree, then in construction disputes the president of the appropriate body, whether the President of the Royal Institute of British Architects, the Institution of Civil Engineers, the Institution of Mechanical Engineers, the Royal Institution of Chartered Surveyors or the Chartered Institute of Arbitrators, may appoint the arbitrator on application of the parties. He will be selected from a panel of expert arbitrators experienced in their particular fields. Very often such arbitrators themselves act as expert witnesses.

Less formality

Another distinct advantage of the arbitration process is that it progresses through its stages in a less formal environment. At the preliminary stages solicitors usually appear before the arbitrator and if counsel's attendance is necessary, he is not robed or wigged. The atmosphere does not have the aura which is sometimes found in courts and which can be traced back to the days of the great jury trials when advocates were renowned for their eloquence or great oratory. Usually there is a sense of what is practicable with a view to the costs involved and an urgency to resolve matters as soon as possible. This puts pressure on the parties to cut down the time for oral hearing and legal submissions to a minimum. However, it is the view of several senior construction arbitrators, including one of the contributors to the *Handbook on Arbitration Practice*, Leslie Alexander, that construction arbitrations are as involved and as prolonged as construction litigation in the Official Referee's court, if not more so.

Procedural stages

For ease of reference and general guidance the expert witness should be aware of the stages in an arbitration so that he knows when he will have to do the greatest amount of work. The expert is usually working regularly on a major case from the date of his instructions but like every process, it has its peaks and troughs.

The stages are:

- identification of matters in dispute – these will form the terms of reference
- appointment of suitably qualified arbitrator
- preliminary hearing for directions
- service of points of claim setting out the claimant's case
- service of points of defence and counterclaim by respondents
- service of reply to defence and to counterclaim
- date for discovery of documents: i.e. the date for exchange of lists of documents in each party's possession which are relevant to the facts in issue in the arbitration and the disclosure of the same for inspection by the other party within a certain period after exchange of lists, usually within seven days
- provision of a Scott Schedule (if necessary) and any other schedules or information which will assist expedition of the hearing
- site visits may be necessary by experts and arbitrator
- exchange of expert's reports and proofs of evidence of witnesses of fact; numbers of experts; access for experts to site
- submissions of legal arguments in writing as may be necessary
- directions as to other written representations by the parties' respective counsel
- communications to the arbitrator (this is primarily of concern for solicitors)
- directions as to plans, photographs, figures in correspondence, agreed bundles of evidence and numbering thereof
- transcript for hearing
- evidence and how it shall be given.

The expert will not ordinarily be present at the preliminary hearing but he must be informed by the solicitors as to the timetable that is directed. He should carefully note in his diary the relevant dates for service of pleadings, for example points of claim/defence etc., and the dates for exchange of experts' reports, site visits, exchange of proofs of evidence, experts' meetings and the hearing itself.

Arbitration hearing

Arbitrators see arbitration as an alternative to litigation and do not like to say that it is similar. However, when counsel is involved in arbitration, it can tend to take on the appearance of a High Court trial. Senior arbitrators have stated their concern on more than one occasion about arbitration being conducted in the same way as High Court litigation. Lawyers may be reluctant to change this for natural and understandable reasons.

Litigation lawyers brought up in the practice of Chancery and Queen's Bench are sensitive to the formalities required by the rules of the Supreme Court e.g. time limits, form of pleadings and compliance with the rules. They are also experts on procedure and evidence and give such matters particular attention. Having said that, however, one finds less anxiety about strict formal requirements in arbitration than litigation and less fear that anything that is done may be subject to criticism by the courts. Ironically, perhaps one does find lawyers more keen on experimentation than non-lawyers. The courts have for a long time recognized that they have strict limits in arbitration and are reluctant to interfere in these procedures. The Arbitration Act 1979 restricts the jurisdiction of the courts so far as appeals are concerned. It is only in exceptional cases that the courts will intervene. In construction cases an application for leave to appeal may be made to the Official Referee. In general, leave to appeal may only be given where it is clear to the judge that the arbitrator was obviously wrong in law on the face of the award.

However, there is a need for a note of warning to be sounded and here the authors would echo Leslie Alexander's words in his contribution to the *Handbook of Arbitration Practice* where he says:

> 'Twenty years ago hearings only had a degree of formality necessary to extract evidence from witnesses but today procedure in arbitration hearings has become so akin to that of the High Court that frequently the only difference is that in arbitration counsel do not wear wigs and gowns.'

At the arbitration hearing the claimant's counsel or solicitor or other advocate may open the case. He will then call his witnesses who will be followed by the expert witness. The respondent will cross-examine after the claimant's counsel has examined in chief. Claimant's counsel can re-examine after cross-examination. The arbitrator may also ask some questions but this is very rare. The respondents follow the same procedure for examination-in-chief and re-examination. The respondent's counsel at the close of his case sums up and he is followed by the claimant's counsel who sums up his client's case.

The arbitrator may then decide to hold a site visit and examine what is disputed or if he has already done so, or it is not otherwise appropriate, he will adjourn the proceedings to consider his award. Once his award is published he is *functus officio*, his work is complete, and he cannot act further in the matter. The award can be enforced by the High Court if necessary.

The roles of the expert and arbitrator distinguished

It is important for the expert who may be called upon to settle a dispute which does not go to arbitration or to court to appreciate the distinction between his role and that of an arbitrator. Sometimes the expert, be he surveyor, engineer or architect, will be asked to resolve a potential dispute and in doing so, he should appreciate the distinctive status of his position.

An arbitrator actually determines a dispute after hearing evidence from both sides in a judicial manner. As a matter of public policy he cannot be liable for negligence because he is carrying out a judicial duty. In arriving at a conclusion in the same way as a judge he cannot be liable in negligence for a judicial act, although he can be liable for a negligent administrative act, e.g. failure to notify parties of hearing.

An expert who settles a potential dispute does not hear all the evidence of both sides; he usually makes his own enquiries and investigations, draws his conclusions and gives an opinion to his client. Legally he can be liable for any damages the client may suffer as result of that advice, for example if certain remedial works are undertaken quite unnecessarily.

Two important House of Lords cases clarified the position on liability. In *Sutcliffe* v. *Thackrah and Ors* (1974) the House of Lords ruled that an architect issuing interim certificates does not act as an arbitrator between the parties. Lord Morris clarified the position by stating:

> 'A person will only be an arbitrator or quasi-arbitrator if there is a submission to him either of a specific dispute of present points of difference or defined differences that may in future arise and if there is agreement that his decision will be binding.'

In the case of *Arenson* v. *Arenson* (1977) it was argued that what clothed a real arbitrator with immunity was not only matters of public policy but also the fact that the arbitrator was a type of judge who did not decide the question before him solely by himself. He had to deal with the contentions of the parties, and that was why the concept of a 'quasi-arbitrator' was argued not to be a justifiable proposition in law. It was further argued that *Hedley Byrne* v. *Heller* (1964) (see above) had established that all persons who expressed an opinion which was negligent were liable to persons who were within a relationship recognized by the law. An expert appointed by one side would do well to remember that he cannot be acting in any sense in a judicial, or so called quasi-judicial capacity, for the fundamental reason that he does not hear both sides of the case, nor are the parties bound to accept his advice. Lord Wheatley defined the judicial tests in *Arenson* v. *Arenson* where he said that the tests were:

(1) that there must be a dispute or difference between the parties which they have formulated in some way or another;
(2) the dispute or difference must have been remitted by the parties to the person to resolve in such manner that he is called upon to exercise a judicial function;
(3) where appropriate, the parties must have been provided with an opportunity to present evidence and/or submissions in support of their respective claims in the dispute; and
(4) the parties have agreed to accept his decision.

It must be the case, therefore, that an expert witness does not and cannot act in an arbitral or judicial capacity; he has no immunity from suit; he is liable just as any other professional person under the rule in *Hedley Byrne* v. *Heller* and indeed must exercise that particular standard of care referred to in Chapter 3.

The expert as conciliator, mediator or ajudicator

Conciliation and mediation procedures

In 1990 the Chartered Institute of Arbitrators, after careful consideration and consultation, published its guidelines for conciliation and mediation. Those guidelines provide for parties to resolve their disputes or differences arising out of matters of contract or other legal relationships amicably without recourse to litigation or arbitration. That is the objective and the ideal but it is not necessarily guaranteed.

The British Academy of Experts also provides training for, and maintains a register of, Approved Mediators and the construction industry is also offered a nationwide service by leading construction contract consultants.

Whilst the objective of mediation is to resolve disputes without recourse to litigation or arbitration, this may not always be achieved but experience in the United States has shown that when parties have agreed to try the process the issues have been resolved speedily.

Mediation

Mediation is essentially an attempt by an appointed neutral person to achieve the resolution of a dispute to the mutual satisfaction of the parties and so dispense with the issues that divide them.

Aggrieved parties usually have a relatively clear idea of what solution would satisfy them, but their concept of what solution would be acceptable to the opposing party can only be the subject of conjecture.

When a mediator is appointed it is essential that he has and maintains the mutual respect of all the parties to the dispute and therefore that he remains neutral in all his dealings with the matters in hand. Whilst he will encourage both sides to 'reveal their hand' to him, the mediator must respect the confidentiality of each party's position and only divulge information to the other party when he has been given specific authority to do so.

In order to make progress, the mediator's role is to persuade the parties to take a reasonable approach to the matters in hand and he can encourage this because he is privy to the position held by the opposing party. Using his own expertise to understand the issues, which are the subject of the dispute, and the declared position and aspirations of the parties, the mediator may suggest to the parties separately how he considers agreement can be reached, introducing elements into the 'package' which may be wholly unrelated to the matters in dispute but which would make sense in business terms. Only if a party so agrees does the mediator then put the proposition to the opposing party, who may have alternative suggestions. Such a procedure will have already broken the impasse of polarised parties.

A mediator does not *impose* a solution on the parties in dispute. This is one clear distinction between mediation and arbitration and litigation. The mediator attempts to steer the parties to positions which each will be able to accept. In other words, he assists them in making their own agreement. He is essentially a catalyst enabling polarised parties to come to a sensible resolution of a business problem.

Mediation is conducted in private. The process can be terminated at any time by either party. Cost of the process is in the control of the parties. They are free to appear in person.

Experienced experts who are interested in this area of dispute resolution can obtain further information from the Chartered Institute of Arbitrators, the Centre for Dispute Resolution (CEDR), the Law Society or the British Academy of Experts.

Chartered Institute of Arbitrators – procedure

The Chartered Institute of Arbitrators' rules provide that conciliators or mediators who act under them act in an independent, impartial and just manner. They should take particular account of:

(1) the general circumstances of the case;
(2) the business relationship of the parties;
(3) the parties' wishes;

(4) the need for a speedy and economical settlement;
(5) whether the matter may be disposed of by the use of a documents-only procedure.

The conciliator has additional powers of requiring submissions of further evidence and/or information, further statements of case and reply, furnishing of samples in sale and supply of goods cases, requiring site visits, convening informal hearings and examining parties; interviewing parties separately, seeking legal and experts' technical advice and requiring parties to provide security for costs.

Settlement of disputes or differences by mediation or conciliation

Under the Chartered Institute's rules the conciliator or mediator may give a preliminary view on the dispute or difference referred to him and the parties may submit their own proposals for settlement. The conciliator or mediator may submit his proposals for settlement to the parties for comment. Where a settlement is reached the parties themselves, or through the conciliator or mediator will draw up a settlement agreement. This agreement may be signed and witnessed by the conciliator or mediator. It takes effect as a contract and not as an award.

Does a mediator/conciliator act judicially?

Inevitably the question may arise as to whether a conciliator or mediator acts judicially and, in so doing, whether he is immune from legal liability other than for purely administrative acts.

Applying Lord Wheatley's tests in *Arenson* to the guidelines for conciliation and mediation published by the Chartered Institute of Arbitrators, the following questions may be asked.

(1) Is there a dispute of difference between the parties which they have formulated? Rule 1.1 of the guidelines provides that they will apply where there is such 'dispute or difference arising out of a contract or some other legal relationship'.
(2) Is the dispute or difference remitted to the conciliator or mediator to resolve? Rule 4.1 provides that 'the procedure is

intended to assist the parties to reach an amicable and equitable settlement of their dispute or difference'. In other words it is intended that he will resolve it by finding a solution which is acceptable to both sides.

(3) Have the parties opportunity to present evidence? Rules 4.2 and 4.3 provide that the parties will have ample opportunity to do so.

(4) Have the parties agreed to accept the conciliator's, or mediator's decision? Under the guidelines the rules do not require the parties to agree to accept the conciliator's, or mediator's proposals for settlement. The guidelines simply provide that the conciliator 'may' and the mediator 'will' submit his proposals for settlement. The guidelines also provide that the proceedings will be concluded where for example, legal proceedings are issued: a party withdraws, or parties give notice.

It appears that the fourth limb of Lord Wheatley's test is not satisfied so that a mediator or conciliator does not act in a judicial capacity. Under some rules it also appears that mediators may communicate with one side or the other but neither side necessarily knows what has been said by their opponent to the mediator. Thus, it also appears there is a breach of the rules of natural justice insofar as each side is entitled to know what case there is against them. In the case of mediation there can be no certainty that the party will know.

Rapid resolution procedure by neutral experts

An interesting procedure has been adapted for use by neutral experts from the Institution of Civil Engineers (ICE) Arbitration Procedure (1983).

Under this scheme each party is entitled to set out his case on issues in the form of a file containing:

(1) a statement of the factual findings he seeks;
(2) a report or statement from, and signed by each party-appointed expert upon whom that party relies;
(3) copies of any other documents referred to by each party-appointed expert's report or statement on which the party relies, identifying the origin and date of each document.

Each party is required to deliver copies of the file to the other party and to the neutral expert in the time directed by the expert. After reading the respective cases the expert may view the site of the works and may require either or both parties to submit further evidence. The neutral expert then fixes a date when he meets each side's appointed expert whose reports or statements have been submitted. At the meeting each party-appointed expert can address the neutral expert and put questions to the other appointed expert. The neutral expert directs the meeting so as to ensure that each expert has an adequate opportunity to explain his opinion and comment upon any opposing opinion. No other person is entitled to address the neutral expert or to question any party-appointed expert unless the parties and the neutral expert agree. After the hearing the neutral expert may make and publish an award setting out such details or particulars as may be necessary giving his decision upon the issues questioned.

Proceedings in the High Court: an outline

The most common type of proceedings for the expert in civil cases are cases in the Queen's Bench Division of the High Court.

In the Queen's Bench Division proceedings are governed by guidelines laid down by the Rules Committee of the Supreme Court which are contained in the White Book or Supreme Court Practice. The White Book provides the form and time limits for pleadings. Pleadings are simply statements of fact in writing argued by each party to establish the issues of the case before trial. They include writ and statement of claim, defence and counterclaim, and reply and defence to counterclaim. Lawyers often serve requests for further and better particulars of various pleadings to clarify certain aspects of the other side's case. In response to such requests, the other side will serve further and better particulars.

Official Referees

Who are Official Referees? These are circuit judges of the High Court who sit in London or in the provinces chiefly hearing construction cases but also having jurisdiction under the Judicature Act 1873 to hear

Experts in conciliation, mediation and mini-trials

1. Experts appearing for parties in conciliation.

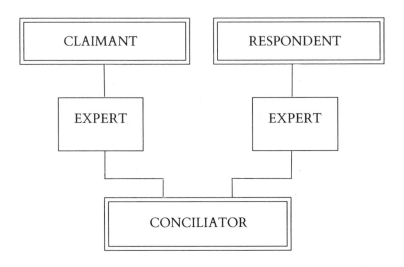

2. Experts given wide powers to resolve in conciliation.

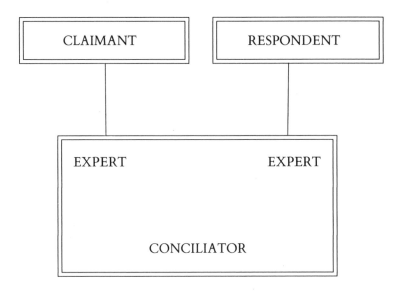

3. Expert as technical assessor in mediation.

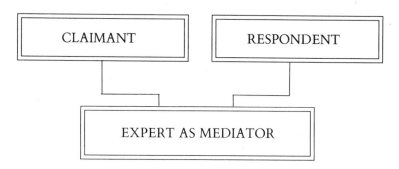

4. Experts as advocates in mediation.

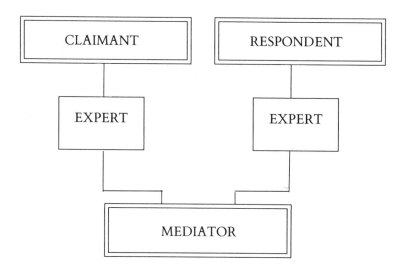

5. Experts in mini-trial.

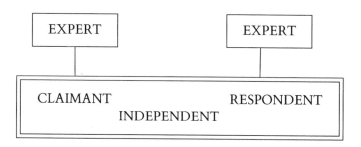

dilapidations cases, or matters involving detail or accounting procedures from all branches of business, commerce and industry. Their function was reaffirmed by the Supreme Court Act 1981. Their numbers have increased to seven in London and several futher appointments have been made in the provinces. Procedurally the work of the Official Referees is governed by Order 36 of the Rules of the Supreme Court 1988.

Over the years it is true to say that their jurisdiction has become more important because of the increasing complexity and financial value of the cases they must hear. At the same time they have developed an experience of coping with voluminous evidence, both written and oral, and have demonstrated a high degree of efficiency in resolving the cases that come before them. Their decisions have also contributed substantially to English contract law and the law of torts.

The procedure is geared to encouraging settlement by narrowing the issues and clarifying the differences before trial. A particular distinction of the Official Referee's court is that the judge deals with all preliminary (known as interlocutory) matters as well as the trial itself, unlike other courts of the Queen's Bench Division where the judges only deal with the trial. Thus parties at trial sometimes have the satisfaction of knowing that the trial judge has been involved from the beginning and will have a grasp of the issues in dispute.

Official Referee's procedure and outline

It will be of assistance to many experts to know in broad terms what the procedure is in the Official Referee's court and how it may differ from arbitration or other litigation. For this purpose they are referred to the standard works on the practice of the Official Referee's court by Judge Edgar Fay QC, *Official Referee's Business*, and a similar work by Judge John Newey QC, *The Official Referee's Court*. Many experts would agree that the procedure is tailor-made for the resolution of construction disputes.

In a building defects case the expert may have been instructed to advise initially on the question of technical responsibility. He will then be asked to advise further during the course of any remedial works and upon the preparation of evidence.

Meanwhile, the solicitors will have issued and served proceedings, including a detailed statement of claim, which will be drafted by counsel, on the basis of the expert's initial report. The statement of claim will be agreed both with the expert and with the solicitor's client prior to service on the other side. The solicitor may have started the case in the Official Referee's court, or may have it transferred there by order of the master from the Queen's Bench Division.

Provided there is no difficulty in any challenge to the plaintiff proceding by way of action in the High Court, the plaintiff's solicitors will then issue and serve a summons for directions, whereby the Official Referee can give directions as to the timetable for service of pleadings, numbers of expert witnesses, duration and a fixed date for trial.

The solicitor will either send a copy of the order on directions or advise the expert as to the nature of those directions so that he can plan his timetable of work for the production of reports and availability for a trial well in advance of those events.

If the expert acts for the plaintiff, he must:

(1) advise on the statement of claim and see that it is factually correct and that it puts forward the technical argument;
(2) advise and comment on the defence and counterclaim so that counsel can formulate a reply and defence to counterclaim and raise requests for further particulars of such pleadings;
(3) assist in preparing a Scott Schedule as required.

If the expert acts on behalf of the defendant, he should:

(1) comment upon the statement of claim so that a request for further and better particulars can be made;
(2) advise and check the defence and counterclaim;
(3) comment upon any reply and defence to the counterclaim and assist with any request for further and better particulars of the pleadings;
(4) assist with replies to the Scott Schedule.

The Scott, or Official Referee's, Schedule, which itemises the issues in dispute between the parties, will be drawn up in the form as directed by the court or as agreed between the parties. Indeed, it would be difficult to conduct the trial of a complex construction case without the assistance of a Scott Schedule. When all the issues in the case have been pleaded by both sides, pleadings are said to be closed.

Following this, a Scott Schedule may be prepared by the plaintiff and served upon the defendant who is required to answer each scheduled allegation pursuant to the Official Referee's directions. Following completion of the Scott Schedule by both sides, the parties will then proceed to discovery. This is an important matter for the expert and lawyers which should be carried out as soon as possible, ideally by both the solicitors and expert working together.

Following completion of discovery, the next direction that will be carried out is the meeting of experts on both sides. Prior to compliance with the formal order for a meeting of experts, the experts may have met to discuss the obtaining of evidence on site, the extent of remedial works required and negotiated in an attempt to agree facts or figures. Prior to this meeting, the experts will have exchanged reports and have a knowledge of the other's opinion. The meeting may then proceed to discuss differences of opinion and the basis for such differences.

Experts' meetings

In an Official Referee's action the usual direction which the Official Referee will give will be:

'Experts of like disciplines to meet without prejudice to try to narrow issues and agree facts by (the Official Referee will fix a date).'

The direction may also state that:

'Experts to seek to agree a joint statement indicating those parts of their evidence on which they are, and those on which they are not, in agreement.'

In *Carnell Computer Technology Ltd* v. *Unipart Group Ltd* (1988), on a summons for directions, orders were made by Judge James Fox-

Andrews QC, Official Referee, not only that the experts should meet, but that they should produce a joint report. The parties agreed terms of reference for the experts' meeting before it took place. At the trial the parties conceded that if the parties had produced a report it would not have been privileged. The Official Referee held that in fact the experts never agreed a report and incidentally remarked:

'The only authority the experts had to reach agreement . . . was vested in them by my directions.

I find that an expert had no implied ostensible authority to agree facts orally or in any form other than in a joint report where an order such as the one I made here exists.'

In *Murray Ltd* v. *UIE Scotland Ltd* (1988) unreported, but referred to in *Richard Roberts* v. *Douglas Smith Stimson* (1989), a case that concerned whether a party should be allowed to call an expert at trial, Judge James Fox-Andrews also remarked of Order 38 Rule 38 of the Rules of the Supreme Court:

'The major objective of the rule is to produce as wide an area of consensus between the experts in which they should indicate in their joint statement those parts of their evidence on which they are agreed. Such agreement will have the effect of removing the protection of "without prejudice" in respect of the matters agreed on.'

In *Richard Roberts* v. *Douglas Smith Stimson* (1989), Judge John Newey QC, Official Referee, said that he respectfully disagreed with the remarks of Judge Fox-Andrews about the authority conferred on experts in agreeing such statements. Judge Newey's view was that: 'a joint written statement made at the conclusion of a without prejudice meeting "without prejudice" is automatically an "open statement".' There were no words in Order 38 Rule 38 of the Rules of the Supreme Court which purported to confer upon experts the power to bind anyone. If it did, then since the rule would not be dealing with procedure but with the substantive rights of persons under the law of agency it would, in his view, be ultra vires.

Judge Newey went on to say that he thought it would be quite alien to the role of the expert witness that an expert witness should have an automatic power to bind parties at the conclusion of 'without prejudice' meetings. An 'expert' does not represent a party in the way that a solicitor represents his client; he is principally a witness and his duties are to explain to the court (and no doubt to those who instruct him) technical matters and to give objective 'opinion' evidence. He said:

> 'If every order for a meeting of experts were likely to result either in an agreement disposing of all or part of the case without either a party or its legal advisers being consulted, orders for meetings would be likely to be strongly opposed.'

The solicitor author of this book readily endorses the view of Judge Newey as a matter of practice. It is not for experts in their role as expert witnesses to conclude all the terms of settlement; it is a matter for the expert witness to report back to the client through the solicitor, or in consultation with the solicitor, the extent of agreement reached on facts and possibly opinion, and it is then for the solicitor to negotiate and agree the terms of the settlement (if such is capable of being agreed) following the experts' meeting.

Objective of Order 38 Rule 38 of the Rules of the Supreme Court

The notes to Order 38 Rule 38 at paragraph 38/37 – 39/6 in the Supreme Court Practice refer to the main objective of the rule as being:

> 'To produce as wide an area of consensus between the experts (as possible), in which case, they should indicate in their joint statement those parts of their evidence on which they are agreed. Such agreement will have the effect of removing the protection of "without prejudice" in respect of the matters agreed on.'

In commenting further upon the effect of this rule in *Richard Roberts* v. *Douglas Smith Stimson* (1989), Judge Newey said:

'EXPERT'S PERSONAL POSITION
If of course during or at the conclusion of a meeting an expert is

persuaded of the correctness of another expert's views and states as much in writing or even orally he cannot subsequently give evidence truthfully to the contrary. The only exception to this is when fresh information, such as the result of new laboratory tests, or a growth in scientific knowledge, has caused him again to change his mind.

If, in the absence of any new development, an expert were to give evidence contrary to that which he had agreed, he could not be cross-examined about what had happened at the meeting because it was "without prejudice", but other sanctions might be invoked against him. If he is a member of a profession, complaint might be made to its governing body; further or alternatively if he is a member of an association of consultants or experts he might be reported to it.

In practice a party whose expert witness has at a "without prejudice" meeting held before trial agreed matters which are contrary to that party's case, is in a difficulty. He can either accept as binding upon himself the agreement made by the expert, or he can replace the expert. If a party adopt the latter course, he will have to apply to the court to order a further experts' meeting and will almost certainly be ordered to pay the costs of the first one.

MEETINGS HELD AFTER COMMENCEMENT OF TRIAL

It may be that Rule 38 is so widely expressed that the court could order a meeting of experts even after a trial has begun, but in practice in the Official Referee's courts meetings are ordered on summonses for directions. Once a trial has commenced meetings only take place by agreement of the parties and this is so even if the court has expressed the view that a meeting would be desirable.

Ideally when parties agree to a meeting of their own experts they should also agree expressly whether the meeting is to be "open" or "without prejudice" and what authority (if any) the expert should have. In the absence of express agreement the parties' intentions have to be inferred.

If, however, a meeting is "without prejudice" while an agreement between the experts is not binding on the parties ratified by them, a party who does not wish to accept what his expert has agreed is in a far more difficult position than if the trial has not begun. He would be most unlikely to obtain leave to call a further expert witness at that stage and particularly not if his only reason for applying is disagreement with the opinion of his present expert.

I think that in deciding whether a meeting held after a trial has begun is "without prejudice" and, whether an agreement reached between experts at it is "open" and binding on the parties, the position during earlier meetings ordered by the court may be taken into account, but in respect of the latter the difficulty of a party's position if he is not minded to accept the agreement is much more important. Obviously the facts as a whole have to be considered; just as they have to be in deciding whether the experts actually reached agreement.'

Possibly, a small footnote to add to Judge Newey's comments concerns the expert's professional liability in exceeding the scope of his contractual authority. It is arguable that in such cases there may be a claim against the expert's professional indemnity insurers and/or the expert himself who is liable just like any other professional who exceeds the scope of his terms of appointment, or his authority and the client suffers damage as a direct result. Although, as yet, there is no clear authority on the point, nevertheless by analogy with that line of cases concerning an architect's authority (see *Sharpe* v. *San Paulo Railway* (1873), *Stockport MBC* v. *O'Reilly* (1978)), any damage suffered by the client, i.e. the cost of instructing a new expert, may fall on the first expert. What is unquantifiable is the amount of damage to the client's case that would necessarily result from changing experts in terms of credibility.

It may well be that an expert who has exceeded authority may be forced to seek independent legal advice especially in the current climate of modern commercial litigation.

Agreement of technicalities

The surveyor author is of the view that it is frequently the case that experts' positions are somewhat polarised at the time the first meeting is arranged, at least on the fundamental issues which form the basis of the dispute. Whilst they may not agree as to the cause of defects, damage or injury and even though they may have expressed widely differing views as to the proposed remedies a meeting of experts can still be productive.

In some cases, it is the first occasion that the rationale of the differing positions of technical experts is explained to both parties. The result of this can be either a measure of agreement; a compromise of position or a confirmation of the polarised positions. It is unlikely that full agreement will be reached at a single meeting but this should not be in any way considered failure.

When meetings have resulted in clarification of the technical views in relation to causes of the alleged damage or injury, further progress can be made which will prove to be very cost effective for the experts' clients. A useful document to come out of such a meeting is a schedule of points on which agreement has been reached.

In relation to the areas of difference which still exist, the alternative technical positions should be considered separately on the assumption that, first, one view is upheld by the court, then, secondly, that each of the other views is assumed to be upheld. For each view it is likely that there is a measure of agreement as to either the necessary remedial work and/or the cost of the remedial works proposed by the other experts. Any such agreement reached in a meeting of experts will save time at the hearing when the whole team of lawyers and experts of all disciplines will be present adding to the costs. Even where agreement as to the costs of remedial work cannot be precisely agreed, such a detailed exercise could well identify areas where, although differences in principle still exist, the likely damages awarded would not be significantly different even if they were to be based on the evidence of the opposing parties' experts.

Meetings of experts on occasions bring to light documents or other evidence which is unknown to one of technical experts and is of such a nature that, had he been aware of such documents or evidence at an earlier date, the opinion expressed would have been modified. This is most likely to occur where the expert has been supplied with a 'selection' of documentation by his instructing solicitor. In these days of prolific photocopying there tends to be a plethora of paper even in the simplest of disputes. This leads to the practice of selecting documents either from clients' files or from discovery of the opposing party's files. Where previously unknown data is identified, the expert would be wise to give serious consideration to the effect of the new data on his technical advice and if necessary be prepared to approach his instructing solicitor to inform him of the new situation.

If the expert is under instructions which unduly restrict his action then he needs to make that clear and discuss the position with his client's solicitor. The solicitor should be able to deal with this and obtain any appropriate further instructions from his client.

The purpose of a meeting of experts is not a 'fishing' exercise for one party to obtain additional information without giving too much away. The idea is to provide an opportunity to clarify technical issues and where possible narrow the areas of difference, some of which may have occurred as a result of misunderstanding, or an absence of knowledge of data that is available.

Judge's visit

In construction disputes there is an increasing practice for the trial judge to visit the site especially where it is considered that such a visit would provide a better understanding of the issues. This may happen before trial or at the conclusion of the opening. On such occasions it is usual for the expert witnesses appointed for all parties to be directed to attend. This is not the occasion on which to attempt to present a client's case to the judge. He has made the visit to see for himself rather than to listen to the views of the parties. It is essentially a fact finding exercise. Such a meeting is likely to commence with the visiting judge making a statement along the following lines: 'I will take note of all I see, gentlemen, but will disregard all I hear'.

Pre-trial conference

Following the exchange of experts' reports and meetings between experts, the parties' lawyers will be directed to attend a pre-trial conference so that the lawyers and the judge can discuss what is the most convenient programme and timetable for the hearing.

If there are several major issues, the judge may decide to split up the trial and have a series of sub-trials. An expert witness may be required throughout the conduct of the whole case and this would mean attendance at all the sub-trials. However, if different experts are engaged for specific purposes, then it is quite likely they will not be required to attend the whole case but just particular sub-trials.

In the Official Referee's court, the order of speeches and proceedings is the same as in other divisions of the High Court. Witnesses of fact precede the expert witness and at the end of the plaintiff's witnesses the defendant's counsel opens his case and calls his witnesses and subsequently his expert.

It should be noted that the judge has a discretionary power to hear witnesses of fact first; see *Alpina Zurich Insurance Co.* v. *Bain Clarkson Ltd* (1989); and see Order 72, Rule A17 of the Rules of the Supreme Court.

Other proceedings in the High Court

Apart from acting as an expert witness for a client in the Official Referee's court, an expert may find himself giving evidence in an ordinary Queen's Bench action before a judge of the High Court or in the Commercial Court. The procedure to be followed is broadly similar to that of the Official Referee's court.

However, the procedure does differ in so far as there is no provision for a Scott Schedule, as such. It is common practice, however, for counsel to draft particular Schedules to facilitate judicial consideration of the pleadings. A summons for directions in the Official Referee's court is usually heard at an early stage whereas it is not generally heard until after close of pleadings in Queen's Bench Division actions, although this is not always the case. Interlocutory applications are dealt with by a master in the Queen's Bench Division and not by the trial judge, as in the Official Referee's court. Interlocutory applications in the Commercial Court are dealt with by a judge.

Basically, the plaintiff's solicitor should issue his writ and serve it within four months of issue; the defendant will acknowledge service; the plaintiff should serve his statement of claim within 14 days after acknowledgement; and the defendant should serve a defence 14 days thereafter. Pleadings may then be closed and discovery takes place. A summons for directions is then issued and at its hearing a direction will be given to fix a date for trial. The direction regarding exchange of expert reports will take effect in practice after the exchange of witness statements of fact.

The expert should note the shorter time limits for service of pleadings, although these can be extended either by order of the court or by agreement. Most of the expert's preparation in such cases should be undertaken before the issue of the writ and after the close of pleadings before the trial. Under Order 38 Rule 2A of the Rules of the Supreme Court the court may order that witness statements be exchanged by a specific date. Expert evidence cannot be adduced at trial without leave of the Court.

Since 5 February 1990 the High Court has a discretion of its own motion to order separate trials for liability and quantum.

Proceedings in the Commerical Court

The Commercial Court was established as part of the Queen's Bench Division of the High Court under section 3(1) of the Administration of Justice Act 1970. It deals with commercial actions which are generally defined as:

'any cause arising out of the transactions of merchants and traders, construction of mercantile documents, export and import of merchandise, affreightment, insurance, banking, mercantile agency and mercantile usage.

Flexibility

This court does not have the rigid formalities that many laymen associate with legal procedures. Whilst it is true that counsel appear in wigs and gowns and the judges are robed, nevertheless the procedure itself is less formal than many would expect. The very name of the court suggests that its work is of a commerical nature which cannot help but affect the nature of the proceedings. The judges are prepared to restrict counsel's rights to argue their pleading points and not to insist on strict proof from witnesses when documentary evidence is considered adequate and sufficient.

Commercial Court procedure

The judges of the Commerical Court have published a *Guide to Commercial Court Practice* (1986), which has recently been revised and adopted by the Commercial Court Committee comprising judges of the Commercial Court and the Lord Chief Justice of England. This document, together with Order 72 of the Rules of the Supreme Court, is essential reading for lawyers practising in that court. It is also of interest to the expert engaged in Commercial Court cases.

Specifically the court deals with:

- contracts relating to ships and shipping
- insurance and reinsurance
- banking, negotiable instruments and international credit
- international carriage of goods
- contracts relating to aircraft
- purchase and sale of commodities
- operations of interntional markets and exchanges
- the construction and performance of mercantile contracts
- law and practice of arbitration and questions connected with or arising from commercial arbitration
- any other matter or question of fact or law which is particularly suitble for a decision by a judge of the Commercial Court.

Chapter XV of the *Guide to Commercial Court Practice* deals with the expert's role and duties in that court. The following is a summary of that chapter.

Order 38 Rule 4 of the Rules of the Supreme Court allows the court to limit the expert evidence. Order 38 Rule 36 imposes further restrictions upon adducing expert evidence.

Exchange of expert reports is governed by the standard directions for trial. It is advisable that exchange takes place a reasonable time after the statements of factual witnesses have been exchanged.

A party should attempt to eliminate or reduce expert issues in advance of the trial. Meetings of experts may be ordered but leave will not be given for calling an expert unless the need becomes inevitable. Different orders may be made for different forms of expertise. Expert evidence is given by reference to the expert's report. Supplementary reports, if any, should be exchanged at least three weeks before trial date. In cases of high scientific content consideration should be given to saving time and cost by means of assessors or the appointment of a court expert under Order 40.

Experts acting in this court should be mindful that solicitors are required to tell the court two days before the summons for directions for trial, how many experts will be required, by what date reports can be served and whether there is scope for agreement between experts. Furthermore, two months before the date fixed for trial the solicitor must tell the court whether any further reports will need to be served and which experts are intended to be called. It is therefore vital that the expert is informed of dates for exchange of reports and hearing dates well in advance of the minimum time limits given to the solicitor to certify these matters to the court.

This court is always anxious to assist the parties insofar as it can, and will encourage the parties to resolve the matters without the need for a trial if possible. This court attaches a great degree of importance to agreeing and resolving matters of expert evidence before trial. Furthermore it is always keen to save time and cost at the hearing. In cases of high scientific content consideration is given by the court to saving time and cost by means of assessors (Order 33 Rule 2(c) or the appointment of a court expert (Order 40). This will be of further interest to the reader in the context of suggestions made in chapter 12 of this book regarding court experts.

Proceedings in the county court

The expert may be introduced to litigation by acting as the expert adviser in a county court matter. The procedure here is similar to the High Court but less formal and involves matters of a more minor nature. Cases are allocated according to the criteria of substance, importance and complexity. Generally cases involving sums below £25,000 will be tried in the county court; those involving amounts above £50,000 in the High Court; those involving amounts between £25,000 and £50,000 in either the High Court or the county court according to the relevant criteria and judicial availability.

For example in a major defects case the building owner may be pursuing the contractors and architects in the High Court, while the owner's tenants are claiming against the building owner for damages for nuisance and disturbance due to the remedial works and damage suffered. In such cases application may be made for the transfer of the minor action to the High Court or the hearing may be adjourned awaiting the outcome of the High Court proceedings. This may mean that the expert will have to give evidence twice over but his reports may be different in scope. In the county court it would be convenient to refer to the main report by way of reference and have a special report in respect of each tenant's complaint.

The procedure in the county court takes the following steps. An ordinary summons is issued by the plaintiff together with particulars of claim. If a defence is filed there is a pre-trial review where the District Judge gives directions and the case is set down for trial. If there is no defence, judgment is entered. The order of speeches and calling of witnesses is the same as in the High Court.

Whichever way the case goes, be it in the High Court or county court, the expert will basically follow the same work programme:

- receipt of instructions
- site investigations and preliminary privileged report
- preparation of further reports and additional investigations
- final report and the giving of evidence in court.

The role of the legal/technical team

In major complex construction cases in recent years lawyers have as a matter of necessity, evolved the idea of teams to deal with the documentation and evidence. That approach was dictated by the complexity and extent of some cases. It has long been the practice for counsel, solicitors and experts to liaise in their independent roles in pursuing litigation or arbitration. In smaller or medium range cases, the expert is not consulted so frequently. In a major case, however, his attention will be required constantly. In that sense, he may have to work with solicitors, sometimes in their office, and with others. He will no doubt find that, of necessity, he must delegate certain areas of his work, for example gathering evidence by way of surveys from the site, quantifying the amount of the claim with the assistance of a quantity surveyor, and obtaining other specialist technical advice, for example the chemical composition of the material which may be in dispute.

On the legal side, it will be necessary for the solicitors to have a partner in charge of the team, assisted possibly by one or two other solicitors or perhaps an articled clerk, or other para-legal assistance. Frequently, it will be necessary for the solicitor's team and the expert's team to work together on some aspects. Just how often such a team functions as a cohesive unit depends upon those involved, and the particular way in which the litigation or arbitration is conducted.

If, as a matter of choice, it is decided that a team would be appropriate in the circumstances of the case, then it should have as its objects:

(1) the co-ordination of all fact finding investigations on site;
(2) the examination of all potential witnesses of fact;
(3) regular conferences to discuss issues of evidence and pleadings;
(4) the formulation of the case based on thorough investigation;
(5) the review of negotiations and further tactics to be employed in order to secure a possible early settlement and save costs and time.

It is suggested that such a team must of necessity function under the direction of the solicitors involved in the case. They co-ordinate the investigations and negotiations, and conduct correspondence with the other side's solicitors. All communications are therefore through them, and in turn they keep members of the team informed as to proceedings in court and when reports or further investigations are required. It might be useful for solicitors involved to prepare a draft programme or timetable of the proceedings from, say, the date of directions given for trial up to and including the date of the hearing, so that every member of the team involved knows exactly what he has to do and when. It is advisable that any programme or timetable should be sufficiently flexible to permit or allow for the unexpected, and sometimes, unpredictable event.

When to do remedial work: timing with proceedings

At common law a plaintiff is under a duty to take reasonable steps to mitigate his loss and damages, so that if he fails to mitigate a defendant may plead such a matter as part of his defence. In certain cases statutory undertakers or local authorities, for example, may be bound by statute to repair the damage immediately. Local authorities have a duty to house their tenants and have certain statutory and contractual duties to repair and remedy defects. The longer they remain unremedied the greater the potential number of claims and the quantum of damages.

The expert working for a local authority may come under intense pressure to get the repairs under way very quickly. This may conflict with his duty as an expert where his prime duty is to give expert opinion on the best evidence. Remedial works may destroy the real evidence of defect. So what is the expert caught in this dilemma to do?

Sometimes both sides can agree that they will abide by the results of a pilot study and apportion responsibility for defects on that basis. On the other hand, as is more likely, both sides will not agree to such a pilot study but may agree to monitor the works as they progress.

(1) The expert should therefore make an assessment and report on what remedial works must be undertaken (this may already have been done as part of the preliminary reports).
(2) The expert must get his client's approval; the remedial works contract should go out to tender as quickly as possible.
(3) The solicitors on both sides should agree monitoring and investigation procedures together with their respective experts.
(4) Arrangements should be made for the collation and gathering of evidence for the purposes of reports, negotiation and trial.

The expert may decide it is useful to have a resident clerk of works or suitably qualified and experienced operative from his office on site all the time to monitor, note and photograph the works in progress. It is important to have a clerk of works on site in whom all parties have confidence.

So far as the court or arbitrator is concerned, remedial works should be completed by the date of the trial or hearing, because only then can

Summary of procedure

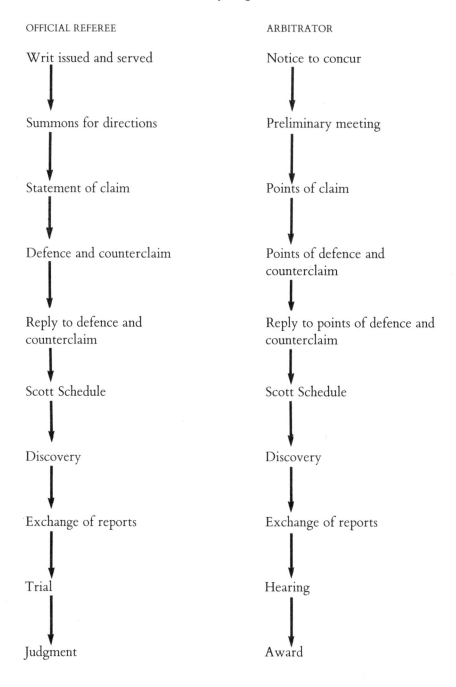

OFFICIAL REFEREE	ARBITRATOR
Writ issued and served	Notice to concur
Summons for directions	Preliminary meeting
Statement of claim	Points of claim
Defence and counterclaim	Points of defence and counterclaim
Reply to defence and counterclaim	Reply to points of defence and counterclaim
Scott Schedule	Scott Schedule
Discovery	Discovery
Exchange of reports	Exchange of reports
Trial	Hearing
Judgment	Award

a full assessment of damages be made. If this is not possible, then the parties' experts may be able to agree the exact nature and scope of the works required.

Checklist on procedure

(1) What type of procedure will be followed in the action? Will it be by way of arbitration or by litigation in a court of law?

(2) If it is to be an arbitration, is it a domestic or an international arbitration? Where will the arbitration be held, which law will govern the dispute under the contract and which law will govern the procedure which applies to the arbitration?

(3) If it is an action in court, in which jurisdiction will the trial take place? Will it be an action in the county court or in the High Court? If the latter, will it be a matter given to the special jurisdiction of the Official Referee?

(4) What is the timetable directed or agreed for the exchange of expert evidence and is there facility for meetings of experts to agree facts? What directions have been given with regard to the mode of trial? Will the trial be split into various sections and when will the various experts be giving their evidence?

7 Evidence and the expert

Introduction

Many building cases are simply disputes as to facts. What actually happened to cause the defect, why did it happen, when did it happen and what was the extent of the damage?

The expert must not only answer these questions but also give his opinion on responsibility from his technical analysis of the evidence.

He must therefore have an understanding and appreciation of the general principles of evidence and show how best he can present his report based on that evidence.

He should have a working knowledge of the different categories and classifications of evidence so as to be able to distinguish which type of evidence is to be used.

Facts in issue

As we have seen in the earlier chapter on investigation of defects and categorisation of issues in dispute, it is essential for the expert to know what precise facts are in issue (i.e. in dispute) between the respective parties. By referring to the pleadings in the case (that is, to the statement or points of claim in the action or arbitration the defence and counterclaim and any reply and defence to the counterclaim as well as to the Scott Schedule), the expert will have a very good idea of what facts are disputed and what he will be required to give evidence about.

Relevance

The expert must address himself to the key question of what facts are relevant to the opinion he is to give. What is technically relevant may not necessarily be legally relevant; the evidence on which an opinion is given may be closely studied by counsel to see that there will be no problems over its admissibility. Matters such as direct or real evidence may be relevant. Indirect or circumstantial evidence may not be. *Phipson on Evidence*, 14th edn, Sweet and Maxwell, defines direct evidence as meaning:

'the existence of a given thing or fact (which) is proved either by actual production, or by testimony, or admissible declaration of someone who has himself perceived it'.

Indirect evidence is defined in *Phipson* as:

'other facts thus proved, from which the existence of the given fact may be logically inferred.'

Where, for example, an architect expert is asked to advise whether the defendant architect was negligent in any aspect of his design or advice to his client, the architect expert will only be in a position to advise on such an allegation once he has made a thorough investigation of the building in question, examined all the drawings and contract documentation, site instructions, clerk of works' reports, and all other contemporaneous evidence, and possibly subjected certain building materials to scientific analysis and obtained results. The expert architect might well then give an opinion to the court that the architect defendant had departed from the standard of care imposed on him at common law by doing something that fell below the recognized practices and procedures (for example, in the use of untried and untested materials). His opinion would be drawn from evidence: the building is real evidence; the drawings direct evidence of fact, as are the contemporaneous correspondence, records, minutes and reports. In addition he might draw the indirect evidence, such as results of tests on certain materials; all such direct or indirect evidence is relevant to the question of whether or not the defendant architect was negligent or not.

All facts discovered and disclosed may not be relevant. The expert must decide on technical grounds. He will be legally advised by counsel and must be mindful of the facts in issue – the facts on which allegations or claims have been formulated, and the issues of defence, counterclaim and reply. All these points will be relevant in the proceedings and must be dealt with as thoroughly as possible.

Other facts may be relevant to 'facts in issue' and these must also be investigated. For example, where an architect advises the use of a certain material which fails because, as it happens in most such cases, the material is incompatible with the subsystem, research and testing of this material would be undertaken and those findings would be relevant to the 'facts in issue'. In such a situation each side would take its own samples and have them independently analysed or otherwise agree a sample and have it tested by one laboratory and agree to abide by their results.

The evidence collated by the expert may include statements of witnesses of fact and other experts. All this will be examined by the expert and his opinion will be sought.

In general the expert must review all the evidence, whether relevant or not to the facts in issue, categorise and collate that evidence, and be ready to refer to irrelevant evidence if necessary – for example, evidence produced by the other side which is not materially relevant to the issue.

Hearsay evidence (as to which, see below) should be avoided unless it is possible to give a statement of information and belief with sources and grounds.

It may be necessary to trace the source of an explanation from several witnesses – who gave the first explanation, is it the best source? It is of importance for the expert to make his own judgment on the basis of the evidence – what he considers technically relevant and why.

Types of Evidence

Direct evidence

Direct evidence is the actual evidence of a fact, for example, the object in issue such as the building. This may be provided by a witness of fact stating that he actually inspected the building in question and saw the actual problems associated with it or, for example, by the writer of a letter who gives evidence as to its content and thereby proves admissibility. The evidence must be something the witness personally experienced.

The best evidence

This is the best evidence that can be obtained. It is direct or real and is usually the production of the real document, piece of material or article in question. It is sometimes known as primary evidence and carries more weight than, say, indirect evidence or circumstantial evidence.

The object of providing the best evidence is:

(1) to establish the basic facts from investigation
(2) to give an opinion based on those facts.

Real evidence

The real evidence is the object itself which is the subject matter of the dispute. Real evidence can be investigated either by on-the-spot investigation on a site visit of a defective building and sample testing, or by acquiring the actual object itself, for example, a defective machine or section of a faulty curtain wall. Every effort must be taken to ensure that the evidence is not tampered with after the defect/accident has occurred, so that there will be no dispute as to the actual state of the object itself.

As soon as possible after a dispute has arisen and both sides have appointed their experts, it is essential that the experts take a joint view of the site or subject matter in dispute. Such experts should be given reasonable notice and facilities for inspection. Obviously, the sooner such inspection takes place, with the possibility of pending remedial works, the better. Photographic evidence should be taken as soon as possible and it is advisable to take a video picture of the whole site in question. Care should be taken as to the quality of video recording. These can often be very poor. Still photographs are clearer and can be taken by the expert himself. Thus, as experts may be helped by a site visit and the gathering of evidence, so too might the judge or arbitrator be assisted in seeing the site for himself at an early date when the defects can be opened up and readily seen. Visual inspection is far more effective than a video or a still photograph.

Sometimes an expert may be asked to report on an alleged defect on the basis of the other side's expert report, but on inspection the expert may find:

- a complete lack of direct evidence
- an entirely different set of circumstances
- or an entirely plausible explanation for the state of affairs.

One must therefore approach the *locus in quo* (site of investigation) with the same circumpection as a detective might approach the scene of an accident or crime.

In examining the real evidence, one must:

(1) Examine and investigate the full extent of the defects/problem and their likely cause or causes. This must be correctly established in order to establish and apportion blame fairly, and to specify the correct remedial work.
(2) Consider the direct and indirect effect and consequences.
(3) Consider legal implications in conjunction with the client's solicitors and/or counsel to establish whether there is evidence of breach of duty in contract or in tort or both.
(4) Take sufficient detailed notes and photographs to draft an initial report and advise the client on appropriate remedial works to mitigate loss.
(5) Make sure that those allegedly responsible for defective work and/or design are given reasonable notice so that they can inspect.

The expert must advise the client of the considerable expense that must necessarily be incurred in order to investigate the case thoroughly. The investigation work must entail opening up problem areas of the works, and in many cases such opening up work and investigation will form the basis of a separate building contract. Apart from investigation the contractor investigating may be required to remedy the problem and consequently there will be a remedial works element in the contract as well. The investigation work can be expensive depending on the nature of the problem but fees in the region of £50,000 to £100,000 are common in major building cases.

Conclusive evidence

Conclusive evidence is the most convincing evidence which is decisive in providing a fact or issue.

Extrinsic evidence

Extrinsic evidence usually consists of oral evidence given in connection with written documents and drawn from a source outside these documents to explain the point in issue.

Judicial view as to the expert's reliance on other evidence

In *English Exporters (London) Ltd* v. *Eldonwall Ltd* (1973) Mr Justice Megarry said:

'The opinion of the expert witness is none the worse because it is in part derived from the matters of which he could give no direct evidence.'

In *H.* v. *Schering Chemicals* (1983) Mr Justice Bingham said:

'If an expert refers to the results of research published by reputable authority in a reputable journal the court would, I think, ordinarily regard these results as supporting inferences fairly to be drawn from them unless or until a different approach was shown to be proper.'

It may also be said, following Mr Justice Bingham's remarks in that case, that an expert holding himself out as having a particular or specialist knowledge or skill in a certain area of his profession should have the appropriate and requisite material to give his opinion and his advice. If he gives an opinion or advice without consulting the appropriate specialist references, then he himself may well find that he has failed to meet the duty of care imposed upon him and failed to advise to the proper standard of an expert holding himself out as having that specialist skill or knowledge.

Indirect Evidence

Indirect evidence can be either mere hearsay or circumstantial. Circumstantial evidence is factual evidence that is not actually in issue but is legally relevant to a fact in issue by reason of its relation to, or in connection with, the matter in question.

For example, the clerk of works may have given a statement to the effect that he recalls a site meeting at which the architect produced a drawing illustrating a particular detail. When the minutes of the meeting are disclosed, there is no direct evidence, i.e. no evidence in writing, on the minutes which state that the drawing was produced by the architect at that particular site meeting. In this hypothetical case no oral or other corroborative evidence is forthcoming from any other person who was present at the site meeting. In such a case the clerk of work's statement is regarded as hearsay evidence.

Circumstantial evidence is indirect evidence, for example, statements from persons who may have suffered similar problems or experiences which relate to the issues in dispute.

Documentary evidence

Section 4(1) of Civil Evidence Act 1968 provides that a statement contained in a document shall be admissible as evidence of any facts stated therein of which direct oral evidence would be admissible if the document:

'is or forms part of a record compiled by a person acting under a duty from information which was supplied by a person (whether acting under a duty or not) who had, or may reasonably be supposed to have had, personal knowledge of matters dealt with in that information and which if not supplied by that person to the compiler of the record directly, was supplied by him to the compiler of the record indirectly through one or more intermediaries each acting under a duty.'

Mr Justice Bingham, after defining section 4 in *H*. v. *Schering Chemicals* said that the intention of section 4 was to:

'admit in evidence records which an historian would regard as original or primary sources, that is documents which either gave effect to a transaction itself or which contain a contemporaneous register of information supplied by those with direct knowledge of the facts.'

Rule 3 of Order 38 of the Rules of the Supreme Court generally provides that evidence produced at trial may be:

(1) by statement on oath of information or belief;
(2) by the production of documents or entries in books;
(3) by copies of documents or entries in books; or
(4) in the case of a fact which is, or was, a matter of common knowledge either generally or in a particular district, by the production of a specified newspaper which contains a statement of the fact.

Prima facie evidence

This is evidence which will establish whether or not the defendant has a case to answer. When a case, claim or allegations are presented by the expert's client, he will usually have submitted the evidence to his solicitor. The solicitor may decide that there appears to be a case for the potential defendant to answer. If there is, he or counsel will advise the client that there is a sufficient question to be addressed to a technical expert. The expert must then decide from a technical viewpoint whether there has been a breach of duty by virtue of a failure to perform to an adequate and reasonable standard – that is, to the standard that the particular profession demands.

Can an expert give evidence of fact and opinion?

The expert can give direct oral or written evidence of his opinion based on his findings of fact. It is clear that he can give direct relevant evidence of fact, e.g. the results of his findings on the opening up work. He can also comment by way of giving his opinion on published statistics and articles but must be wary of commenting upon hearsay. See *English Exporters (London) Ltd* v. *Eldonwall Ltd* (1973).

In *Harmony Shipping Co.* v. *Saudi Europe* (1979) Lord Denning said:

'The court is entitled in order to ascertain the truth to have the actual facts which he has observed adduced before it and to have his independent opinion on these facts.'

The court is entitled to have the expert's independent opinion on the documentary evidence before it. The expert's evidence on matters of fact stands in the same position as evidence of a witness of fact. Factual evidence already adduced before a court or tribunal may be commented on by the expert if required. On the other hand in cross-examination the expert may be asked to assume certain facts and then state his opinion. He should be most careful to qualify his answers in such a way as to distinguish supposition from facts.

In a Practice Note given by Mr Justice Ackner (as he then was) following the decision in *Ollett* v. *Bristol Aerojet Limited* (1979), guidance was given as to what must be substantially disclosed by way of exchange of reports under Order 38 Rule 38(1). The judge declared that the expert had a duty to report on the factual description of the subject of the action, e.g. a defective machine, the circumstances of the accident or problem (the defect itself), and the expert's opinion on such matters. The correct sequence of the report for exchange would therefore be to:

(1) establish the basic facts from investigation;
(2) give an opinion based on those facts.

Affidavit evidence (written evidence on oath)

An affidavit is a written statement of evidence sworn by the person making it (the deponent) before a person authorized to take affidavits, and where admissible, receivable in legal proceedings as evidence either in support of an application, or in answer or in reply.

It is rare that an expert witness will be called upon to give evidence on affidavit, but there may be occasions when it will be necessary for him to swear one. This may occur, for example, if there was some necessity during the intervening or interlocutory stages; or if, say, he was acting for a defendant client in Order 14 proceedings where a plaintiff had applied for summary judgment – the expert may be in a position to say that as a quantity surveyor (for example) he believes there is a strong defence to the plaintiff's claim, and he can then give figures on affidavit to prove that belief. Again, the expert may be unavoidably out of the country at the time of the trial, and in such a case, the court may accept his evidence on affidavit. Affidavit evidence does not carry as much weight as oral evidence at the hearing because the witness's testimony has not been subject to cross-examination.

If swearing an affidavit, the expert should note that affidavit evidence is not admissible at trial without leave of the court. If the court does give leave, then the witness may be cross-examined. If the witness fails to attend and be cross-examined, his affidavit may be excluded as evidence.

A deponent may give a solemn affirmation instead of taking the oath required.

The expert can give his opinion on admitted or assumed facts; indeed this is the reason for his giving evidence.

In general an affidavit may only contain such facts as the deponent is able to prove of his own knowledge; Order 41, Rule 5(1) of the Rules of the Supreme Court. There are however, the following exceptions:

(1) where it is used in interlocutory proceedings the affidavit may contain statements of information or belief with the sources or grounds thereof (hearsay);
(2) an affidavit filed in support of or against summary judgment under Order 14 (see above) may contain statements of information or belief with the sources or grounds thereof (hearsay);
(3) the court may order evidence on affidavit of any particular facts;
(4) second hand hearsay may be admissible provided that a deponent specifies in his affidavit the name of the person who could give

direct evidence of the fact in question; Order 41 Rule 5(3).
(Second hand hearsay may be defined as 'hearsay-upon-hearsay',
e.g. where A the expert gives as evidence a statement that C
made to B and B related to A.)

Oral evidence given on oath

Section 16 of the Evidence Act 1851 provides that every court, judge,
justice, officer, commissioner, arbitrator or other person lawfully
authorized to hear, receive and examine evidence is empowered to
administer an oath to all witnesses that are legally called before them.
An oath is defined as 'a religious asseveration by which a party calls his
God to witness that what he says is the truth or that what he promises
to do he will do.'

The oath

The form of oath is administered by an officer of the court. The witness
holds the New Testament (the Old Testament for a follower of the
Jewish religion or the Koran for a Muslim) in his uplifted hand and
repeats:

'I swear by Almighty God that the evidence I shall give shall be the
truth, the whole truth and nothing but the truth.'

In the case of a person who is not a Christian, Jew or Muslim, the oath
is administered in any manner which is lawful: section 1, Oaths Act
1978.

Under section 5 of the Oaths Act 1978, any person who objects to
being sworn may be permitted to take an affirmation instead. If it is not
reasonably practical without inconvenience or delay to administer the
oath in the appropriate manner, the witness may be required to affirm.
The words 'oath' and 'affidavit' include 'affirmation' and 'declaration';
'swear' includes 'affirm' and 'declare'.

The affirmation

The form of affirmation is:

'I [name] do solemnly, sincerely and truly declare and affirm that the evidence I shall give shall be the truth, the whole truth and nothing but the truth.'

So far as arbitration is concerned, section 12(2) of the Arbitration Act 1950 provides that unless the arbitration agreement expressly provides otherwise, an arbitrator can examine, on oath or affirmation, the parties and witnesses in the reference.

He does not necessarily have to take evidence on oath. In rare cases, the court may order such evidence to be given by affidavit: section 12(6)(c) Arbitration Act 1950.

The legal significance of the oath is that, if a person breaks that oath by making a statement in the course of the proceedings from the witness box which is material and which he knows to be false or does not believe to be true, he commits perjury. If he manufactures false evidence with the intention to deceive and mislead, that is an offence at common law.

Hearsay evidence

In investigating matters in issue, the expert witness must be careful to distinguish hearsay from other evidence and understand what exceptions are made to admit such evidence. As a general rule, any oral or written statements made by a person or persons other than the witness who is giving evidence are hearsay and inadmissible to prove the truth of the facts stated by the witness. The hearsay rule only applies where the rules of evidence govern the proceeding. The reasons for this rule are various: the person making the statement to the witness was not under oath; he was not observed in court and was not subjected to testing by cross-examination; and the statement was not made in public.

Exceptions to the rule against hearsay evidence
So far as the expert witness is concerned there are certain exceptional circumstances that may arise where the courts are prepared to admit hearsay evidence. Some of these exceptions are as follows.

(1) Hearsay evidence of opinion is admissible under section 1 of the

Civil Evidence Act 1972 which extends the operation of sections 1, 2, 3 and 4 of the Civil Evidence Act 1968.

(2) Section 1(1) of the Civil Evidence Act 1968 enables an expert to give evidence of a statement made to him by another person.

(3) A statement made by a person whether orally or in writing may be admissible in evidence of any facts stated therein of which direct oral evidence would be admissible; section 2 of the Civil Evidence Act 1968.

(4) Section 2(5) of that Act provides that if a statement is not made in a document it must be proved by the person who made it, heard it, or otherwise perceived it.

(5) Section 4 of the Civil Evidence Act 1968 (as amended by section 1(2) of the Civil Evidence Act 1972) provides that an opinion contained in a record (document) will be admissible in proceedings provided the evidence of the person who originally supplied the information is also admissible. Section 1(2) admits an expert's opinion based on official records, e.g. governmental or institutional publication.

(6) Statements taken from computer printouts are admissible under section 5 of the 1968 Act provided the computer was in proper working order and the information was obtained under the ordinary and proper course of its usual working. The expert must be careful in using such evidence and if he has any doubts must seek the advice of counsel or a solicitor.

(7) Section 2(1) of the Civil Evidence Act 1972 provided that restrictions and procedures imposed upon the calling of witnesses need not apply in the case of experts who placed reliance upon their reports for the purpose of giving their evidence. Experts do not need to give statements. Where an expert produces a report and he is not being called, then reliance may be placed upon section 4(1) of the Civil Evidence Act 1968 whereby the expert statements contained in that report may be admitted as evidence of any fact therein of which direct oral evidence would be admissible.

The section then goes on to describe the document's status. It is especially important for experts. The expert must have actually inspected the real evidence, e.g. a building, or he must have relied upon a 'record' compiled by a person who had personal direct knowledge of that evidence. For example, an expert architect may rely for particular matters of detail upon a junior assistant from his office who had inspected the real evidence, i.e. made a survey of the building and listed its defects. It will be admissible if, say, the architect asked his junior for a number of surveys of defects and the junior in turn delegated that task to an independent surveyor who acted under the instructions of the expert's junior. In that way the indirect evidence of the surveyor may be admitted in the report provided the expert explains how this evidence was obtained at his junior's instruction. The architect could give evidence of the surveyor's report which may be classified as 'double hearsay', but if that evidence was critical then the lawyers would undoubtedly advise that the surveyor should be called to give evidence.

Under the rules provided by virtue of section 4 of the Civil Evidence Act 1968 the Rules of the Supreme Court enable experts in the High Court to refer to articles and statistics. Plans, photographs and models are generally admissable in evidence provided the other side has been given the opportunity of inspection and agreeing their admission. Inspection must be given at least 10 days before trial. In complex building disputes an arbitrator will give specific directions on this evidence if required.

The dangers of arguing hearsay

Arguing a point in issue on the basis of what you have been told by somebody of no particular qualification is dangerous and often inadmissible as mere hearsay. This is illustrated by the case of *English Exporters (London)* v. *Eldonwall Ltd* (1973), which involved an expert valuer. Mr Justice Megarry, as he then was, said in that case that an expert cannot simply retell what others have told him because the other side are entitled to have a witness of fact 'whom they can cross-examine on oath as to the reliability of the facts deposed to and not merely as to the witness's opinion as to the reliability of information that was given to him not on oath and possibly, in the circumstances, tending to inaccuracies and slips'. Mr Justice Megarry further described the practical difficulty which was presented to the court thus:

'It is often difficult enough for the courts to ascertain the true facts from witnesses giving direct evidence without the added complication of attempts to evaluate a witness's opinion of the reliability, care and thoroughness of some informant who has supplied the witness with the facts he is seeking to recount.'

What the expert must do is to prove the case by admissible evidence and take heed of the warning against hearsay laid down by Mr Justice Megarry. He expressed the view that an expert valuer witness may:

(1) Express opinions that he has formed as to values, even though substantial contribution to the formation of these opinions has been made by matters of which he has no first hand knowledge.
(2) Give evidence as to the details of any transactions within his personal knowledge in order to establish them as matters of fact.
(3) Express his opinion as to the significance of any transactions which are, or will be, proved by inadmissible evidence (whether or not given by him) in relation to the valuation with which he is concerned.
(4) Not give hearsay evidence relating to details of transactions not within his personal knowledge in order to establish them as matters of fact.

Generally the expert should observe the guidelines and the warning, confining his reports to giving an opinion on matters of fact known directly or indirectly to him as a result either of his own observations or of his general professional knowledge or experience.

If the expert refers to another witness's evidence that evidence must have been admitted as a statement of fact by that witness. Normally it will apply where a statement is exchanged in accordance with the order on directions. The statement of evidence will have been admitted and accepted by the other side as evidence. It also applies where the expert has a written statement from the party for whom he acts and has been told that the witness will be called to prove it. This is one of the reasons why the expert is frequently called last, or at least after certain witnesses of fact. If the witness does not come up to proof the expert can then often rather hurriedly reassess his opinion upon the evidence of fact as it has actually been given.

Who can give evidence?

As a general rule in civil proceedings, all normal adult persons are competent and compellable to give evidence. The criteria are whether the witness is able to understand the proceedings and the nature and effect of the oath, and whether he is able to speak the truth and give evidence in a rational manner.

What is proof?

Proof is the establishment of the existence or non-existence of some fact to the satisfaction of the court. The means of such proof is the material evidence i.e. how facts are proved.

The burden of proof

The burden of proof has two aspects – legal or persuasive and evidential. The former is the burden to prove each particular allegation against the defendant. That burden lies with the plaintiff in civil proceedings and with a claimant in arbitration matters. It is for him to prove the truth of his allegations on the balance of probability – to prove that his interpretation of the facts is the most probable and believable one.

The expert has a key part to play in this role by explaining how this is probable and why it is to be believed. What if the expert has a doubt? An expert may give evidence of his opinion even if he has a doubt, provided he makes this clear in his evidence. The question is whether *on balance* he is satisfied that his opinion is the best. A cautious expert is often more impressive than a smug, self-satisfied person who sees no difficulty in a difficult matter.

The standard of proof

It has been said that the burden of proving a case is on the party making the allegations. The standard of proof required is proof on the balance of probabilities.

For example, if a claimant has direct photographic evidence of each defective piece of tile on a wall and the respondent has only a limited and restricted sample of the tiles, the court is more likely to conclude on the balance of probabilities that the tiles were defective as shown on the photographic evidence produced by the claimant.

Admissibility of the expert's report

The expert's report will only be admissible if:

(1) leave is given to adduce expert evidence or the parties agree;
(2) the trial judge admits it as evidence;
(3) it contains matters in issue or a degree of relevance to the facts in issue.

Lawyers sometimes refer to documents being 'admitted as a document'. This simply means that the document is classified as evidence. It is either genuine or a true copy, but that does not mean to say its contents are true. This must be proved. Likewise the expert's report and the opinion it gives, although admitted as a document is subject to proof by examination or agreement. The judge will decide what is and what is not admissible as opinion evidence.

Evidence relating to another witness

It was held in *S and B (Minors)* (1990) that an expert pyschiatric witness could give an opinion of another witness's mental state, but could not express an opinion as to the truthfulness of that witness's evidence.

Examination of Site

In the process of examination of the site the expert will rely on physical inspection, what he actually saw, and what he observed, the taking and testing of samples, and the statements of witnesses of fact. Statements of witnesses of fact must be supplied by the solicitor in charge of the case. He should speak to all those involved directly. Interviews with clerks of works, architects, quantity surveyors, the builder's operatives and others directly involved may be useful once the expert has carried out his initial investigation.

Any such discussions the experts may have on site as a matter of evidence are classified as hearsay and inadmissible. The expert must inform the solicitor of any such conversations so that the solicitor and counsel can decide whether proofs are necessary and whether any particular person should be called as a witness. What other witnesses of fact may say may have a bearing on the expert's opinion.

Evidence of damage

Where legal liability is to be proved, your client's legal advisers will advise on what is required. The main point to bear in mind is that there must be proof of legal and/or physical damage which must have occurred as a direct result of a breach of duty in tort or contract which was reasonably foreseeable. If there is no damage, there can be no loss and no claim at law.

One must always ask: what is the actual damage to this building now or in the reasonably foreseeable future? Is the damage directly related to the breach of duty? Is there adequate, sufficient and admissible evidence of the breach and the damage? If there is no actual evidence of damage, will damage occur in the foreseeable future? Could damage be reasonably expected to occur and for what reasons? Is it possible to quantify the damage in financial terms?

Photographic evidence

Photographic and, more recently, video tape evidence can be most useful in building up a comprehensive picture of what the subject matter in dispute actually looked like at the material time a course of action arose. Contemporaneous photographs are essential. The expert should exercise his judgment as to the precise number required but should err on the side of taking too many rather than too few.

Some experts may be satisfied with a sample of photgraphs, while an opponent can always counter with more photographs not restricted to the limits of the sampling exercise. Photographs, video tapes, film and tape recordings are all admissible as evidence, but perhaps the still photograph can give more definition to a particular defect than a moving video picture. On the other hand the moving picture can place the subject in proper context by panning the area.

A checklist of points on evidence

The expert should bear in mind the following points when considering evidence:

(1) What evidence will you produce?
(2) Is it all in your report?
(3) Is it admissible?
(4) Has counsel advised on evidence – have you raised any doubts about the evidence?

8 Discovery

What is discovery?

In simple terms discovery is your client's opportunity to see what evidence is in the possession of the other party to the proceedings. It is a legal process whereby documentary evidence in the power, custody or possession of a party to an action or an arbitration is disclosed to the opposing party or parties to the proceedings. 'Documents' that can be discovered include recorded material, e.g. tape recordings, video and computer recordings. Discovery takes place when pleadings are closed pursuant to the directions of the judge or arbitrator as the case may be.

Importance

Under the rules of court it must be emphasized that discovery of documents is the job of the solicitor; it is not the job of the experts. There are cases however where, as a matter of practicality, a solicitor may ask the expert to assist with the process and it is only in that particular respect that the expert will be directly concerned with this process.

Indirectly the expert will be affected by discovery and interested to assess the technical importance of documents in the other side's possession. These documents may be vital insofar as they may affect his analysis of the facts and his opinion on them.

Discovery comes before the service of any Scott Schedule. There may be many documents that have a direct bearing on the replies required for the Schedule and hence the necessity of the expert seeing such relevant evidence that may affect technical replies and subsequently the facts and the opinions of his final report.

Assisting with discovery

Whether the expert is advising in a High Court action or an arbitration matter the orders and directions given for discovery have a similar effect and meaning. In a building dispute this can mean a substantial amount of pre-contract, contract and post-contract documentation emanating from the architect, the employer, developer, contractor and sub-contractors, engineer, project manager, management contractor, works contractors, quantity surveyor, valuer, solicitors and other consultants. All the relevant documentation originating from them is disclosable unless it is privileged from production. It must be emphasized that discovery is principally the task of the solicitor and the expert can only assist with it if expressly instructed to do so. The extent of the expert's involvement must be clearly defined by the solicitor in each case.

If he is to assist with discovery the expert must know his own side's strengths and weaknesses and decide what evidence he needs to obtain or should be looking for in the other side's documentation. He will want to discover and investigate the documentation not only for trial purposes but also for the purpose of obtaining evidence of facts which he may be able to agree or not and/or evidence of facts which may counter the other side's expert's conclusion.

The solicitors will have examined the other side's documentation to ascertain what evidence supports their client's case and what evidence is against it. The expert will be able to assess the technical importance of any contradictory evidence and deal with it in his report.

The following suggestions are made as to the expert's likely role in discovery subject to express instructions from the legal team. He may:

(1) request a detailed brief for discovery from the solicitors, giving clear indication of the type of technical issues they require the expert to examine. That may either be a general or particular and detailed brief depending on the type of case;

(2) obtain a copy of the relevant pleadings e.g. statement of claim and defence and seek explanations of any allegation or counter-allegation that is not understood;

(3) attend with the solicitor at discovery and obtain copies of any document through the solicitor that may be relevant to the issues of the evidence upon which the expert or his opponent may seek to rely for proving his report and conclusions, i.e. any material likely to give assistance to either side's experts;

(4) review such evidence in (3) to see whether there is a basis either for agreeing facts with the other side's experts or possibly producing a joint report to save costs. The result of such review should be discussed with the client's solicitors and instructions taken before there is any meeting of experts;

(5) test the basis for the expert's opinion derived from analysis of his client's evidence. If the basis is fundamentally undermined so that the expert's opinion can no longer be sustained, the expert should say so. If evidence is produced which puts certain points in doubt, again the expert should tell the solicitors and state such reservations in his report to the solicitors' clients.

As a result of disclosure of any evidence by the other side the expert may be put on enquiry and asked by the solicitors to give views on particular matters. Such tasks will necessarily be indeterminable and entirely dependent on the type of case and evidence which may well generate considerable additonal work.

Discovery may necessitate further tests and investigations and in some cases be the subject of a report in itself.

Non-disclosable evidence

Evidence disclosed upon discovery should not be 'privileged' or of such nature as it would be unconscionable to disclose for reasons of public policy.

What is privileged?

Privilege is a very difficult legal subject and this book can only give the expert the merest outline of the types of evidence that may by their nature be privileged. In each particular case where a question of disclosure and admissibility of evidence may arise, it is a matter for the lawyers to advise the expert and the client, not for the expert witness. Because the expert witness has a particular role and of necessity must work closely with the legal team, it would be wrong if he did not have some idea of what is privileged and what is not. If there is any doubt legal advice must be sought.

'Privilege' is the legal term that applies to non-disclosable evidence, for example communications between solicitor and client, 'without prejudice' correspondence (that is, correspondence which genuinely attempts to settle or agree issues in dispute) and notes of any such conversations and meetings, advice and opinion of counsel, and expert advice and opinion given strictly for the purposes of considering, or in contemplation of proceedings.

An example of privilege is the expert's initial report which may be provided at the request of the client's solicitor to enable him to consider with counsel the adequacy and sufficiency of facts from which counsel and solicitors conclude that there may be a prima facie case for the defendant to answer.

Reports

If the expert has compiled a report on technical responsibilities in the certain knowledge that his client was considering legal action, or if he prepared it at the direction of his client's solicitors prior to the issue of proceedings, and proceedings were contemplated, then that report – if indeed that were the *dominant purpose of* the report – is privileged from disclosure. This may also cover any correspondence with the client or solicitor upon the same criteria. In *Waugh* v. *British Railways Board* (1980) the House of Lords ruled that a document would only be privileged if it had been brought into existence for the 'dominant purpose' of litigation. This would seem a very difficult test to pass and it is by no means certain that every expert's preliminary or other report will pass the test.

Proofs of evidence

Privilege extends to witnesses' proofs of evidence unless the privilege is mutually waived. Official Referees find advantage in such disclosure in cutting court time on examination-in-chief. Under Order 38 of the Rules of the Supreme Court, the High Court has powers to order the exchange of witnesses' proofs of evidence. (See Chapter 4.)

Communications between solicitor and expert witness
These are privileged and cannot be communicated to the court except with the consent of the party concerned. Lord Denning in *Harmony Shipping Co* v. *Saudi Europe* (1979) said that:

> 'If questions were asked about it then it would be the duty of the judge to protect the witness (and he would) by disallowing any questions which infringe the rule about legal professional privilege.'

Waiver of privilege – documents prepared for civil proceedings produced in criminal proceedings

It was recently held in *British Coal Corporation* v. *Rye (Dennis) (No. 2)* (1988) that there was no waiver of privilege where documents prepared for the purposes of civil proceedings were made available in criminal proceedings.

In that case the plaintiff sued the defendant to recover approximately £2M damages which it was contended had been overpaid to the defendant in connection with works for the repair of buildings and land damaged by the plaintiff's mining activities. Prior to the commencement of the action, the plaintiff instructed a firm of quantity surveyors to investigate the works carried out by the defendant and the cost of the works. A number of reports were prepared by surveyors. As a result criminal charges were brought against the defendant. The reports were disclosed to the defendant as part of the advance disclosure of the prosecution's case. In the course of that trial evidence was given by one of the plaintiff's employees. He had a large number of documents including reports from the surveyor, and various witness statements prepared for the purposes of civil proceedings. The trial judge ordered the prosecution to hand over those documents connected with the criminal charges to the defendant, who was acquitted. The plaintiff sought to recover the documents disclosed to the defendant in the course of the criminal proceedings. The defendant refused. The plaintiff obtained an order in the civil proceedings for their return and an order that the defendant make no use of them or information derived from them in the civil proceedings.

It was held, dismissing the defendant's appeal, that when the documents were brought into existence they were protected by legal profesional privilege. The plaintiff's actions in making the documents available for the criminal proceedings did not constitute a waiver of that privilege. It was contrary to public policy for the disclosure of privileged documents in criminal proceedings to constitute a waiver of the privilege attached to the documents in civil proceedings.

Privilege – 'without prejudice' correspondence and disclosure to third parties.

As a general rule, 'without prejudice' negotiations which lead to a settlement between parties to the action or dispute cannot thereafter be disclosed to a third party. They remain privileged.

In *Rush and Tompkins* v. *Greater London Council* (1988) the plaintiffs entered into a building contract with the first defendants and engaged the second defendants as sub-contractors for certain work. The plaintiffs began an action against both defendants, but the plaintiffs and the first defendants settled their claim after an exchange of 'without prejudice' correspondence and negotiation. The second defendants then applied for discovery of this correspondence, but the plaintiffs claimed that it was privileged. The judge held that their claim was valid and dismissed the application. The Court of Appeal allowed the second defendant's appeal and ordered discovery, but the Greater London Council appealed to the House of Lords and this appeal was upheld.

Their Lordships held that in general the 'without prejudice' rule made inadmissible in any subsequent litigation connected with the same subject matter, proof of any admissions made with a genuine intention of reaching a settlement. Admissions made in order to reach a settlement with a different party in the same action were also inadmissible, whether or not settlement was reached with that party. The general public policy applied to protect genuine negotiations from being admissible in evidence also applied to protect those negotiations from being disclosed to third parties. The judge's decision refusing discovery was therefore restored.

Privileged until disclosure

In *Derby* v. *Weldon* (1990) it was held that an expert witness's report prepared by a party to litigation for the purposes of the trial is privileged until disclosed by that party. If, however, he expressly disclaims any intention to produce oral evidence on the topic referred to in the report, the court cannot override the privilege by ordering disclosure.

In the Court of Appeal, Lord Justice Dillon stated that when a report was disclosed it lost its privilege, but that did not necessarily waive privilege automatically over the thoughts of the expert on a topic with which he had expressly disclaimed dealing. It was not necessary for the expert, for the purposes of putting his client's full case before the court, to anticipate all possible lines of cross-examination that might occur to him. Subsequently it is suggested that whilst the expert must disclose the substance of his report and the whole of his report, that does not require him to frame his report so that he can anticipate all the likely lines of cross-examination to which he will be subjected by the other side. See the discussion on *Kenning* v. *Eve Construction Limited* (1989) and a consideration of the expert's role in experts' meetings in Chapter 12.

Protecting evidence

If there is any danger that documentary evidence may be destroyed, then the client's legal advisers may consider invoking a court's jurisdiction to seize and preserve the evidence by means of an 'Anton Piller order' (*Anton Piller* v. *Manufacturing Processes* (1976)). This is usually exercised in cases involving films, tapes and records, but orders can be made to protect documentation where there is a risk of its removal or destruction: *Emmanuel* v. *Emmanuel* (1982). The importance of the evidence and the risk are factors that must be considered before this exceptional measure is taken.

Preservation of evidence may be secured by the court granting an order. The Rules of the Supreme Court provide that the High Court can order the detention, custody or preservation of any property which is the subject of the cause of matter or as to which any question may arise. The court can also order inspection of such property. Under Order 29 Rule 3 the court can order samples of the subject matter to be taken or order experiments to be tried on or with such property.

Early inspection

The expert should inspect the property in order to satisfy himself as to the nature of the defects and the extent of damage caused to property. In certain, rare circumstances early inspection of property can be ordered by the court but it is usually the case that inspection can be agreed simply through respective solicitors. If there is any danger of evidence being destroyed, then naturally the expert will want to take the earliest opportunity of inspecting the property to see the actual state of the defects before any urgent remedial works are carried out.

It may be a good idea for the expert to suggest to his client's solicitors that the judge or arbitrator inspect the site prior to, or during, the course of the remedial works.

If before discovery there is anything the expert is not sure of relating to matters in issue or evidence, he must clarify these with the legal team. He must fully appreciate what needs to be proved as a matter of fact and what needs to be proved by way of his opinion evidence. For ease of reference and guidance, the expert will need to refer to the points of claim or statement of claim, the subsequent pleadings such as defence and counterclaim, but more particularly he may be assisted by the Scott Schedule which will provide a brief synopsis of the issues in dispute.

If a number of documents are disclosed that the expert considers highly relevant to his client's case and his view on the matter and these undermine the whole or part of the other side's case, then the solicitors advising the client should discuss the importance of such a matter both with the expert and with counsel and decide whether or not the matter can be raised forthwith as an issue and in the negotiations. This will involve the expert in giving his view and support in the negotiations. The expert must judge the weight and strength of the point from a technical standpoint; if he feels it is valid and of significance then the matter must be pressed as hard as possible in the negotiations.

Considerations of costs

One of the main tasks for an expert must be to get the best possible terms of settlement for his client without him incurring too much by way of costs. This must be borne in mind, particularly with the opportunity presented by discovery. Up to the stage of discovery the client will have incurred certain costs in initiating or defending the proceedings, with limited involvement of counsel. After discovery, with preparation of bundles of evidence and other preparations for trial, the costs will escalate considerably and counsel will be involved more frequently.

If in the course of discovery evidence is found that prejudices or undermines the expert's opinion on a technical aspect, or causes his client's lawyers to have serious doubts about putting forward the allegation in court or maintaining it in the pleadings or Scott Schedule, it is better that such a matter be excluded than run the risk of incurring costs in proving it. The expert must be aware that his costs for assisting with discovery may not be allowed unless these costs are incurred (a) with the client's consent and (b) are reasonable in all the circumstances.

In *James Longley* v. *South West Thames Regional Health Authority* (1983) it was held that such costs of expert assistance (not necessarily related to discovery) would be allowed on the basis that such evidence was necessary to form an informed judgment.

In *Reynolds* v. *Maston* (1986) Mr Justice Bingham (as he then was) considered the general practice of expert witnesses who had been warned to attend a hearing but in the event were not called to give evidence and consequently did not receive any cancellation fee for the time set aside and not used. However, in the special circumstances of the case before him the experts were entitled to receive compensation for time set aside where the case had been settled on the day before the date fixed for trial.

It is not usual to allow the expenses of experts who are called not as witnesses but merely attend court to advise counsel (*Consolidated Pneumatic Tool Co.* v. *Ingersoll Sergeant Drill Co.* (1908)).

Quantum element in evidence

'Quantum' is another term for monetary value. In searching the other side's documentation the expert must not only look out for evidence that establishes liability, whether in tort or in contract, but must also look at evidence relating to the quantum element of his client's claim. It is essential to prove that the client has suffered loss as a direct result of a breach of contract or tort by way of physical damage or economic loss. This can be based on the real evidence which the expert may possibly have examined, documentary evidence in possession of both parties, oral evidence to be given at the hearing, but more particularly the expert's opinion based on that evidence.

It may well be that the expert has been furnished with some proofs of evidence or statements obtained from witnesses by the solicitors, and these can then be compared with the evidence obtained in the process of discovery. When discovery is completed, the expert and his client should have a much better and more precise estimate of exactly what quantum of damage is involved and a clear idea of where the responsibility and technical liability lie.

Before the hearing

Again, the expert will need to decide from a technical point of view, for the purpose of writing his report for exchange and preparing for examination in court, what evidence supports his client's case and the basis on which he, the expert, is supporting it, what evidence is relevant to the issues and what is the best evidence. Unless this exercise is carefully carried out, he will not cover all the issues that counsel may require him to deal with for the purposes of the final report. This will basically form the substance of the expert's own oral direct evidence at the trial or hearing.

It is essential that evidence disclosed on discovery should be considered by the expert who may give evidence. He must see what the other side has, because he himself may be called on to give his view on that evidence as well as the evidence put forward by the other side's experts. It is no use delegating this work to an office junior: time and time again discovery may reveal some unexpected finds. These may be quite useful in supporting the client's case.

Solicitors must examine the other side's documents. That is their duty to the court as officers of the court but the expert also has a duty to review the technical evidence in giving his report and examining evidence obtained from discovery. He may well seek to rely upon such evidence in presenting his report and giving evidence orally.

Solicitors are sometimes not content merely to rely upon photocopy documentation as a substitute for inspection, albeit that the process of an inspection is a somewhat ancient process. Original documents may be required and certainly solicitors examine those very carefully especially where such originals refer to documents that are unavailable or unobtainable. The solicitors may not be satisfied that discovery is complete and they may issue a summons requiring the other side to verify their list of documents by way of affidavit.

Useful hints

In complex technical cases the expert may attend with the solicitor as suggested. If so he should be responsible for making his own notes as to what documents he may require for the purposes of reference and what copies he may require for inclusion in his report. In reading through the evidence disclosed the expert may consider a number of points that have not been previously addressed, and the evidence may also raise further technical questions that he will wish to address. The expert may also seek to make further enquiry or investigation. Discovery is a process that may last several days depending on the volume of documentation. It is essential that if the expert does attend on discovery he keeps within the bounds of relevance and that, although he may be intensely interested in a particular technical aspect or subject, he must always have an eye upon the particular issue in the proceedings and the reason why he is being called to give evidence.

So far as technicalities and scientific proof are concerned, counsel will rely upon the expert's view of the matter. Whether such evidence he may give will be admissible will be a matter for the judge but at this stage the expert must concern himself with relevance to the issues as pleaded. The context in which the expert refers to disclosed evidence on discovery will be a matter for the expert to consider in the context of his report and the technical subject addressed.

Completing discovery

Once discovery has been completed either by the solicitors or with the assistance of the expert, the expert will want to consider very carefully whether or not his opinion has changed in the light of new evidence. If his opinion has changed he must say so and if necessary write an appropriate letter or report to the solicitors. They will then consider whether or not that necessitates changing any particular pleading or allegation, or indeed whether further investigation must be made to the case. By the end of discovery the expert should be able to give reasoned and objective opinion to this client and solicitors as to whether, upon the basis of the technical evidence, the case appears sufficient or not. If not, he must be frank.

Apart from the documentation a clear indication of the client's real position may come from proofs of evidence that have been obtained by the client's solicitors. Where there is exchange of proofs of evidence between the respective parties' solicitors, experts will review those proofs in the context of the evidence disclosed.

Summary of the purpose of discovery for the expert

So far as the expert is concerned, discovery is a critical stage of the legal proceedings. His aim in assisting with this process should be:

(1) To obtain evidence, to substantiate and corroborate his views and generally support his client's technical and legal case.
(2) To assess the strengths and weaknesses of his client's case and the opposition's case and to reassess it in the light of disclosed evidence. (This may result in amendments to pleadings and in the joining of other parties in the action.)
(3) To clarify uncertainties and narrow the issues in dispute.
(4) To assist in negotiations and agreeing matters of fact in order to save costs and court time, using the evidence to the best possible advantage of the client.
(5) To obtain further evidence for final report.
(6) To ensure that he has seen all the relevant evidence relating to the issues in the case before giving evidence at a hearing.

(7) To assist generally the legal team with technical opinion on matters of fact.

(8) To assist the court or arbitrator in giving evidence in an honest, clear, professional and concise manner.

Discovery and evidence checklist

In dealing with evidence obtained on discovery and in evaluating such evidence an expert should bear in mind the following matters.

(1) What are the issues in the case?

(2) Have pleadings been placed at your disposal and have you had the benefit of understanding all the points in issue and have these been explained?

(3) What is your client required to prove on each issue?

(4) What evidence do you rely on for your opinion?

(5) What evidence supports your client's case?

(6) What evidence is against your client?

(7) How do you assess and evaluate (5) and (6)?

(8) Are you able to agree any evidence/conclusions with your opposite number?

(9) Has discovery brought the parties closer together or put them wider apart?

(10) Could you refer to any of the evidence obtained from the other side on discovery?

(11) Have you any doubts about your opinion on any issue upon discovery and exchange of witness statements? If you have such doubts, have you told solicitors and counsel? Are they fully aware of any such doubts?

(12) Are you confident that you can support your opinion/opinions in your report on the basis of all the evidence?

9 Preparation of Scott Schedules

Introduction

One of the directions that may be given in a construction case is the direction that:

> 'the plaintiff (or defendant) serve on the defendant (plaintiff) a Scott Schedule (in the form to be agreed between the parties' solicitors, or in default of agreement determined by the court) (form to be determined by the court) by [a certain date to be agreed or fixed].'

The object of such schedule is to dispense with the constant reference to volumes of pleadings. Sometimes the schedule is referred to as an 'Official Referee Schedule' but it is usually described after its inventor. Those interested in the history of this form of pleading in the Official Referee's court are referred to the work by Judge Edgar Fay QC, *Official Referee's Business*, and to a similar work, on the practice of the court, by Judge John Newey QC.

The type and form of the schedule will vary according to the particular case. The judge may give specific directions as to the column headings or accept those as drafted and agreed between the respective parties and their counsel. The schedule has a great advantage over pleadings for the expert: firstly, because it is in a form he will technically understand at a glance; and secondly, because it does away with the constant cross-referencing of pleadings and further and better particulars. All the claims and allegations, counterclaims and basis for such should be within the confines of the schedule.

The schedule has also been adopted by arbitrators and they may give directions for the production of such as a matter of practice.

What type of schedule?

The type of schedule depends on the nature of the case. If it is a defects case, it will usually take the form of a schedule of defects and damages. Sometimes counsel may require that this is drafted in a particular form to suit the particular case, e.g. where particular details of defects are required for each room of every house in a particular block on a estate.

In the main, counsel will require a detailed schedule based on the pleadings he has drafted and a statement of the case as it stands, i.e. omitting any items previously agreed. The items in the schedule should be clearly cross-referenced so that neither the judge nor counsel wastes time in conference or in court ascertaining exactly what is alleged or refuted. Alternatively, counsel may draft the schedule with the assistance of solicitors and the expert.

The pro forma schedule on the following pages for a building defects case is suggested as a mere guide. The expert witness will usually assist counsel in deciding the best and most appropriate form depending on the type and complexity of the case.

The item and description of defects

The item number should refer to the item in order of sequence in the schedule. The description of defects should start with a reference to the allegation of claim in the statement of claim and give a precise description of the defect.

It should state, for example:

'The DPC in the area of the [. . .] to [. . .] was omitted contrary to the specifications [item . . .] and in breach of the Building Regulations [regulation . . .] all contrary to [para . . . of the code of practice 19. .]. The contractor omitted this and by so doing was in breach of Clause [. . .] of the contract dated [. . .]. The architect

Pro forma Scott Schedule

Property/type	Defect element, report reference, Sharp & Co.'s report	Defect	Breach Alleged	Breach of codified terms*	Relevant regulations byelaws, codes of practice	Remedial work done since appointment of Sharp & Co. (respondents) – A1 Nos 242, 286, 248	Cost (£)
(1)	(2)	(3)	(4)	(4a)	(5)	(6)	(7)
1 Windsor Court, 5 River Court	Roof D.1	Faults allowing water penetration or creating a risk thereof and faulty insulation.					
	D.1.1.2, D.1.2.1, D.1.2.2, D.1.3.1, D.1.4.1, D.4.2, D.1.6.1	(1) Missing and inadequate pointing to the stepped flashings at the junction of link bedrooms with the gable end and inadequate wedging to the said flashings.	(1) Negligent *supervision*	(1) AC	(1) Byelaw 50	(1) Removed stepped flashings and insert new flashings in conjunction with new cavity trays (D.1.6.1).	1,728.20
	D.1.1.2, D.1.2.1, D.1.3.1, D.1.4.1, D.1.6.1	(2) No cavity trays provided at the junction of link bedrooms and gable ends.	(2) Negligent *design*	(2) BC	(2) Byelaw 50, Building Research Digest No. 11, Oct. 1959	(2) Install cavity trays (D.1.6.1).	
	D.1.4.1, D.1.6.1	(3) Missing and incomplete pointing to stepped flashings at junctions with adjacent dwellings and inadequate wedging to the said flashings.	(3) Negligent *supervision*	(3) AC	(3) Byelaw 50	(3) Flashings repaired, rewedged and repointed (D.1.6.1).	
	D.1.6.1, D.1.6.3	(4) Ceiling insulation to link bedrooms laid untidily and incompletely.	(4) Negligent *supervision*	(4) A		(4) Provide new insulation to link bedroom ceilings to current standards (D.1.6.3).	

* Breach of Codified Terms: The following code is used hereunder:

A – Breach by respondents of obligation to provide dwellings which were constructed in a good and workmanlike manner.
B – Breach by respondents of obligation to provide dwellings which were constructed of materials which were of good quality and reasonably fit for their purpose.
C – Breach by respondents of obligation to provide dwellings constructed so as to be fit for human habitation.

Note: Where more than one letter appears against a defect and no preposition is used, the words 'and/or' should be read in between each letter.

Remedial work carried out at claimant's expense before appointment of Sharpe & Co.	Cost (£)	Consequential losses	Amount (if any) admitted by respondents (£)	Respondents' observations	Claimants' reply	Respondents' observations on the allegations in columns 4a and 13
(8)	(9)	(10)	(11)	(12)	(13)	(13a)
		Payments to date to tenants rehoused while remedial work was being carried out:		Roof D.1	1 Windsor Court, 5 River Court	1 Windsor Court, 5 River Court
		5 Windsor Court Removal expenses	300.00	The respondents deny that the conditions alleged occurred in each dwelling and/or any condition so found results from negligence on their part and/or that any condition so found has resulted in damage to the structure.	*Note:* Save as hereinafter appears and safe insofar as the same consists of admissions, the claimants join issue with the respondents on Column 12.	Roof D.1
		Loss of rent while remedial work was carried out:				(1) Breach of Codified Terms A and/or C denied.
		1 Windsor Court 5 River Court	11,107.10 3,957.07		Roof D.1	(2) Breach of Codified Terms B and/or C denied.
				Negligent design and/or negligent supervision is denied for the following reasons:	(1) – (2) – (3) – (4) – (5) –	(3) Breach of Codified Terms A and/or C denied.
		Detailed calculations of such rent losses will be submitted within 21 days of service of the schedule.		(1) and (3): That the alleged faults arose as a result of failure to carry out appropriate maintenance. (2): No damage arises.		(4) Breach of Codified Term A denied.
				In response to items (1), (2) and (3) at Column 3 and item (5) at Column 6 the respondents deny the necessity of the works carried out		

failed to insure that the DPC was laid in this area and failed to make an adequate inspection or supervise in accordance with Clause [. . .] of his contract.'

This type of drafting will encompass alleged breaches of contract against the contractor, breaches of the Building Regulations, and hence statutory duty, breach of contract by the architect and breach of his duty of care in negligence. The party alleging must be precise and specific in the wording of the schedule. Although this is not the duty of the expert, it will assist counsel if the expert appreciates the arguments that can turn on the precise wording of a description of a particular defect: e.g. if the schedule says 'widespread on all roofs'. It is better that each defect is precisely located on summary sheets appended to the schedule, if possible.

The damages claim

The damages that can be claimed by a plaintiff under a building contract are measured by the cost required to rectify defective work to bring the building up to the standard that was contracted for, i.e. the state in which the building would have been had the contract been performed properly in accordance with its provisions. In certain cases it will not be appropriate to claim damages because the cost of remedial works would outweigh the likelihood of recovery of such cost from the other party. In such a case the plaintiff may claim that he has suffered a 'diminution in value' which is assessed by a valuer as a certain percentage of the current market value of the property at the date of the hearing.

The defendant's comments

The defendant here refers to the particular paragraph of his defence which deals with the item of claim. He may make a straightforward denial by passing responsibility for the defect to, say, the architect. If a contractor, he might say that he obeyed the architect's specific instruction on the matter and omitted a certain detail. The architect, in his turn, may deny and defend by saying that he redesigned the detail after his client protested it was an extravagant waste of money and he could not afford it. On the other hand a contractor may admit he built a certain particular detail badly but may dispute the cost of reinstatement, in which case he will admit the claim but argue that the value of reinstatement is less than the claim.

Official Referee's/arbitrator's comments

A column is left for the use of the trial judge or arbitrator.

The expert's role

In some cases the expert witness will draft a Scott Schedule for the approval of the legal team. This happens in cases where the expert has had some experience of the type of case before and has given evidence in court on previous occasions. The expert who is acting as expert witness for the first time can assist counsel by putting forward draft schedules but would do better to be guided by what counsel and solicitors advise.

There is no general rule about drafting the schedule. Some lawyers will require very full description and detail, while others may require only a very concise statement. Both will suffice for counsel's purpose because in the final analysis it is counsel who has the responsibility of pleading the case before the court and arguing it against opposing counsel. If there are any uncertainties or ambiguities in the draft, these must be eradicated to the satisfaction of the expert and counsel.

The schedule will be used by counsel to conduct the hearing and may be used by the experts to discuss aspects of the case during 'without prejudice' negotiations. The advantage of a Scott Schedule is that it states the issues and consequently assists in narrowing the differences for the purpose of such negotiations.

On considering the draft schedule the expert and lawyers may consider some items insignificant – the pursuit of such items could result in throwing away the costs, or costs could exceed the damages recovered. Some items in the schedule may be agreed as defects subject to the determination of the issue of liability.

Drafting the schedule

If the expert is asked to submit a draft to counsel he should base the schedule on the form perscribed in the judge's directions and the advice of counsel as to headings. He will base it on the statement of claim if acting for the plaintiff and any reply to the defence and counterclaim. Further and better particulars should also be reviewed to see whether any additional allegations emerge. Having compiled the list of allegations, the expert should refer to his report to see what the allegation is based on and state it concisely. He should state the cost with support, where necessary, from a quantity surveyor and/or valuer.

Counsel will check the draft and revise the wording appropriately. The expert may cross-reference the pleadings to his report. In the report he will refer to codes of practice, Agrément Certificates, trade association codes and to the Building Regulations, depending on which regulations or codes were in force at the time when the damage occurred. Once the plaintiff's case has been completed, the schedule is served on each defendant and their comments are entered. When each party has completed it and served it, the schedule is regarded as a summary of the pleadings and used accordingly by counsel at the hearing.

Apart from the formal Scott Schedule in major building defect cases, other schedules may be of considerable assistance to the court to illustrate clearly where, for example, a contractor has been in delay or where an architect has been late in issuing instructions. Such schedules can also be of value in loss and expense claims and in arbitration generally.

The great advantage of scheduling is that issues are clearly analysed. There is no wastage of words. The expert should make use of his natural and professional skills, be he architect, quantity surveyor or building surveyor, to illustrate his opinion and views by sketches, drawings, graphical displays and schedules. Computers and word processors may facilitate good presentation. However, the expert is encouraged not to overdo it.

Checklist for preparing a Scott Schedule

(1) Who is to draft the schedule?
(2) What matters have to be covered in the schedule?
(3) What are the headings of the schedule?
(4) How much detail is required for each item in the schedule?
(5) What other schedules may be appended to the Scott Schedule for further detail and description purposes?
(6) Are there any other schedules or devices by which the case could be presented clearly in an effort to save costs and time?

It is always open to the parties' counsel during the course of proceedings or in trial to suggest ways and means to the court by which time and cost could be saved. Experts may be actively encouraged to assist the court in this respect.

10 Preparing for trial: the final report

Once the pleadings are closed, the Scott Schedule drafted and served, and discovery completed, the expert should concentrate all his efforts on preparing his final report. This report will comprise the basis for his oral evidence in court. It will also be the basis for final negotiations and the subject of a response by the other side's expert after exchange. The basic objective is that the expert should state the facts and findings of his investigations and give his opinion and conclusions on such facts.

Legal status of final report

Prime objectives

The aims of the rules of court in this context are to encourage settlement and agreement on as many issues of fact as possible. Accordingly the rules of court provide for disclosure and exchange by the parties of their respective experts' reports. (See discussion on Order 38 in Chapter 4.)

Other objectives

The rules relating to exchange are also aimed at:

(1) avoiding surprises at trial and avoiding lengthy argument of matters of fact;
(2) obviating the need for experts' attendance at trial where matters

can be agreed, so shortening the evidence and consequently the costs of the case.;

(3) enabling the respective experts to prepare their arguments more carefully and thoroughly.

Overriding these objectives, the court aims to ensure that equal opportunity is given to presenting a case adequately. The rules seek to prevent a party from obtaining another party's report without having disclosed his own. They also ensure that both sides have equal opportunity to call their own expert or experts.

Restrictions

Expert evidence, and hence the report, can only be produced in evidence at trial if:

(1) the court gives leave
(2) all parties are agreed
(3) the court directs.

Contents of a final report

Ideally the report should contain:

(1) The expert's qualifications and experience in this field. It should state on whose behalf the report is produced and the date of the expert's instructions.
(2) A description of the brief and what the expert was instructed to investigate. It should detail the layout of the report and index, and information on which the report is based. It should identify sources and list authorities.
(3) A description of the location of the estate/property, ideally by sketch maps, elevation drawings, photographs, plans, etc. and brief background history.
(4) Each item of claim or the response to claim should be detailed as per the Scott Schedule.

(5) Comments on the defects or matters in issue giving a clear opinion.
(6) Comments on the extent of damage and remedial works carried out or to be done.
(7) The cost of repairs quantified.
(8) A description of how the expert arrived at his conclusions and on what evidence.
(9) Reference to the evidence to support the opinions, description, conclusions etc.

Subsequently, when the expert has studied the other side's reports, he should give his views on them to his client's solicitors and counsel. Having given such comments the expert may be requested to produce a further report.

The opinion must follow the facts

In writing his report the expert must follow the golden rule that the opinion must follow the facts. He must set out the facts of the case, describe events, circumstances, or findings, and then give his conclusions and opinion. He must do that in respect of each particular event or investigation. It is noteworthy that in Australia there is a rule of evidence at common law that, except in straight-forward cases where the facts are admitted and readily identified, the opinion of an expert is admissible *only* where the premises, i.e. the facts upon which his or her opinion is based, are expressly stated (*Trade Practices Commission* v. *Arnotts* (1990).

Report for the plaintiff/claimant

If the expert is acting on behalf of a plaintiff in a court action or claimant in an arbitration he must carefully consider the principles of the law of evidence outlined in Chapter 7 and the duty which he has to discharge with regard to the burden of proof. The burden is of course on the plaintiff/claimant to prove his case on the balance of probability. In other words the expert's explanation of cause and failure must be presented in such a manner as to convince the court that his explanation of events is the most probable and likely in all the circumstances. He will only be able to do this if he has carefully considered the technical evidence and investigated the probable causes of failure thoroughly. There is no substitute for thorough investigation and meticulous recording of evidence to substantiate conclusions.

In a relatively small case or arbitration matter of under, say, £50,000 in value, a report may be fairly straightforward and based on a limited sample of defective units. This approach may also be used where the parties have agreed that a pilot study or its finds can be used as the basis for deciding liability and the scope of remedial work. If the parties can agree that liability should be apportioned accordingly, all well and good, but if a party objects, the case will be decided in the ordinary way on the weight of the actual evidence.

In a small case a report may contain in each section:

(1) photographs of the defect;
(2) reproductions of drawings – may be reduced to A4 size for convenience of report or as A3 folded;
(3) findings on the preliminary site investigations;
(4) findings on further detailed site investigations;
(5) causes and reasons for failure;
(6) reference to legislative and consultative documentation;
(7) description of remedial works carried out;
(8) cost of remedial works/repairs; diminution in value.

In a larger construction case, for example defects on a major housing development, the report is a major undertaking and will take months of preparation. It is a labour-intensive operation but well worth the effort so far as the client is concerned if success in proving the case is achieved.

A major report should follow generally the guidelines set out under the heading 'Contents of a final report', above, but to be more particular they may contain the following:

(1) the brief
(2) the layout of the report, i.e. how the report is constructed and how it is intended to be read – what each volume contains
(3) the information on which the report is based

(4) the introduction
(5) the defects
(6) test results
(7) remedial works
(8) costs.

Within this framework will be:

(1) the layout of the report
(2) the index
(3) the site plan
(4) plan and cross-sections of various details in question
(5) location plan
(6) the brief from the client.

The information upon which the report is based

This should refer to all those documents etc. that the expert has consulted and on which his opinion is based, for example: the architect's drawings, the specification, bills of quantity, correspondence, statements, orders, instructions, invoices, manufacturers instructions, site dairies, weekly reports, monthly returns, day work sheets, site investigations and surveys, previous reports, scientific records and research laboratory tests, results of consultation with specialists, information from sub-contractors, their correspondence and records, and advice from quantity surveyors and technical officers.

The introduction

This should present a short history of the building or development in issue and how the problems arose.

Defects

This section should provide a description of each defect, its investigation and how this was carried out, the results, photgraphs, site plans, tests and results and a discussion of the causes of failure and technical responsibility for failure. Location plans should be included showing the precise location of the problem areas.

Specialist's report

The main theme of a specialist's report is that of cause and responsiblity. The expert must give his opinion on the probable cause and his reasons for diagnosing the cause. He must also be clear as to who is responsible professionally. It may be one party or body, or it may be several. He must say who is responsible and why, and whether or not, in his view, the defendant in question has been negligent – whether he has fallen below the standard one would reasonably expect from a member of his profession or calling.

The lawyers will form their own view on the basis of what the expert says about technical responsibility. A breach of a code of practice, for example, is some indication that the appropriate standard has not been followed and possibly the standard of care was not maintained. A catalogue of design failures and bad workmanship will also tend to suggest evidence of negligence. These are matters for the legal advisers to consider in conjunction with the experts. It should be remembered that the judge or arbitrator will decide the issue on the weight of evidence and arguments.

Report for the defendant/respondent

It is always difficult for a defendant's expert to respond to a case without knowing precisely what the other side's expert is going to say. In Official Referees' cases the judges have wisely tried to prevent games of surprises in inconsistent reports by the use of the Scott Schedule. By this device and by exchange of reports, both sides' experts know the other side's arguments and can easily follow each item of defects as tabulated in the schedule.

Tactically the defendant's expert must seek to disprove, if possible, the allegations against his client on technical grounds. He may seek to do this by:

(1) producing different evidence and/or evidence of counterclaim;

(2) demonstrating there is no damage as a result of the alleged defects;

(3) arguing that there is less damage than is claimed.

He may also contest the matter by giving his view that the plaintiff's damages should be drastically reduced by reason of betterment; for example, where the remedial work substituted superior materials for inferior ones at much greater cost than the original material.

The defence expert's report may therefore be along the following lines:

(1) list of contents
(2) brief
(3) background discussion
(4) allegations in the Scott Schedule
(5) alleged extent of the problem
(6) opinion on alleged defects
(7) documentation references, for example to bills and drawings
(8) discussion of allegations
(9) comments on allegations and plaintiff's expert report
(10) evidence of condition of the property concerned
(11) damage
(12) comments on remedial works and extent of betterment.

Once the report has been written, the expert will submit it to the legal advisers who should not take any part in actually writing that report, but they can always advise the expert on matters of law and evidence just as the expert can advise the lawyers on technical matters. If the expert has definite views on matters and has stated these in his report, he should be cautioned about changing such an opinion. He should not be tempted in any way to say things simply to please his client or the lawyers and he should, as a matter of duty, resist any attempt to change his evidence for the sake of tactics. There is a grave danger that if he does so, he will regret it when he stands in the witness box under cross-examination, having sworn to tell the truth, the whole truth and nothing but the truth.

If the lawyers advise that certain allegations are vulnerable because the evidence is doubtful, then it is best to drop such matters and concentrate on what can legally be proved in evidence.

Presentation of the final report

The report should be easy to read and written in plain, clear language. Jargon must be avoided and all technical terms should be simply explained.

There should be logical progression so that each section leads clearly to the next. Each section should be headed. There must be a clear line of reasoning in describing the defect and in drawing technical reasons for failure. It should be clear on what evidence you have based your opinions. There must not be ambiguity or vagueness.

Wherever possible the report should simply refer to a schedule of location of defects or plan of location – this is simple, precise and far more convincing.

The report should be divided into sections along the lines described earlier in this chapter. Shorter rather than longer sentences are preferable, although sometimes it will be necessary to be precise and make specific qualifications. Diagrams, sketches, plans and simple calculations are always of assistance provided there are not too many to confuse.

As to the physical presentation, the report should be properly bound and titled with the name of the firm and the individual who will be giving evidence. The individual responsible should sign the report and date it. The report should be paginated to avoid confusion and should be properly indexed with a list of contents. Photographs should be identified and dated. A description of what is illustrated is essential. The report must be comprehensive with as little need to refer to other documents as possible. Remember that this document will be referred to in the witness box, and it is better to have one document rather than numerous others and various cross-references which may confuse the judge and counsel. The main narrative may be broken up into various issues, ideally in a comprehensive manner, so that each may be dealt with comprehensively.

The report, the whole report and nothing but the report

All of an expert witness's report should be disclosed for the purposes of evidence given in court, not merely the bare bones.

The rule in *Ollett* v. *Bristol Aerojet* (1979) was followed in *Kenning* v. *Eve Construction* (1989) where Michael Wright QC (Deputy High Court Judge as he then was) decided that, where the defendants' expert engineer sent a report dealing with allegations of negligence in the statement of claim to the defendants' advantage, but also enclosed a covering letter stating possible causes of the accident which involved negligence on the part of the defendants:

(1) since the points raised in the expert's letter were fairly obvious and it was not inconceivable, especially after careful study of the photographs, that the plaintiff would have sought leave to amend the statement of claim even if the letter had not been available for perusal; and that in those circumstances the plaintiff would be granted leave to amend; and

(2) although the letter was a privileged document, the defendants had a duty, under Order 25 Rule 8(1)b of the Rules of the Supreme Court, to disclose the substance of an expert's evidence to be given at the trial; that that disclosure included not only evidence to be given in examination-in-chief but matters that could arise in cross-examination; that, accordingly, the defendants would have the choice of either not calling the expert to give evidence at the trial and retaining the privilege attached to the letter or calling the expert and disclosing the substance of the letter; and that, therefore, the court would make no order on the defendants' application as it was a matter for them how they exercised their choice.

This whole point seems to have emerged as a result of one of those exciting, unpredictable occurrences that happen every so often in the course of litigation. The defendants' solicitors in the action decided that they would call their expert at the trial of the action to support their client's case. They disclosed voluntarily and unilaterally a copy of their expert's report. No expert's report was disclosed on behalf of the plaintiff. By some inadvertence a junior member of the defendants' solicitors' firm sent off not merely the expert's report but also the covering letter from the expert.

As many of those involved in personal injury actions know, it was the practice before Mr Justice Ackner's Practice Direction in *Ollett* v. *Bristol Aerojet Limited* (1979) to have two reports; one report for the purposes of disclosure and exchange, the other report for the benefit of the solicitors. That practice was denounced by Mr Justice Ackner because he required that for the purposes of the Rules of the Supreme Court the substance of the evidence must be disclosed. In the case of *Kenning* v. *Eve* (1989) it was clear to Michael Wright QC that only part of the expert's opinion had been disclosed and that the expert's covering letter to the solicitors, which gave an entirely different complexion to the report, was a fundamental part of that opinion. The solicitors then had the choice either of calling that expert to give his full opinion, or not calling him at all and seeking another expert's view. They decided to disclose the report, and once they had decided that, they were under a duty to disclose the totality of the expert's opinion and not just part of it.

The question was in this case, as in most cases, 'was the substance of the evidence communicated to the other party?' It would be useful for solicitors to bear in mind the following guidance given by Michael Wright QC:

'As I say, it seems to me that the solicitor's choice is simple. He must make up his mind whether he wishes to rely upon that expert, having balanced the good parts of the report against the bad parts. If he decides that on balance the expert is worth calling, then he must call him on the basis of all the evidence that he can give, not merely the evidence that he can give under examination-in-chief, taking the good with the bad together. If, on the other hand, the view that the solicitor forms is that it is too dangerous to call that expert, and he does not wish to disclose that part of his report, then the proper course is that that expert cannot be called at all.'

The expert's letter to the solicitors was a privileged document which would not ordinarily, under the Rules, have been disclosed. However, in Michael Wright's view, if the solicitors wished to call their expert then, because the letter directly related to the substance of the report and the expert's opinion, the letter would have to be disclosed. If it was not, then the expert's view would undoubtedly have been disclosed on cross-examination and his report discredited.

The case raises the spectre of the ethics of experts and their evidence – to what extent should an expert insist that his total view be given in the report produced for exchange? Is an expert liable if he does not state the whole truth in his report? Is he in any way liable for any professional misconduct in circumstances where he produces a report which does not deal with the weaknesses as well as the strengths of the case?

The answer to these questions may be that it depends upon how the expert perceives his duty, the nature of his briefing from the solicitors, the advice given to him by counsel prior to preparing the report and the expert's own assessment of how much he can leave unsaid. In view of the comments earlier in this book concerning the 'hired gun' approach, it is felt that it is extremely difficult for an expert to produce a report that is unacceptable to the legal team. The expert has no right of appeal to his own professional organization; all he can ultimately do is to make his position very clear when giving his evidence in court to the judge. If it is a very difficult point and a point which the expert cannot accept with any amount of persuasion from counsel, then the expert has a duty to tell counsel that it is a point that he cannot support, will not support and, if counsel presses him, he will have to resign his appointment. That is an extreme case and one which no counsel will really press because counsel would obviously be aware of any hostility in the witness and would advise the solicitors to seek a different opinion.

At the same time, a judicial mood can be detected whereby judges are tired, quite understandably, of seeing the same experts produce reports which take up entrenched positions, whilst the judges know full well that in court the experts will concede certain points.

Possibly time might be saved and points of conflict in opinion evidence resolved earlier, if the court intervened before trial. The Official Referee's court might be the forum, for it is already the forum for several experiments, including computerisation which as been pioneered by Judge Bowsher QC.

Is there a case for a Special Referee?

The practice is that once experts' reports are exchanged, copies of the reports are lodged in the Official Referee's court with the judge's clerk. This is a formality and it is unlikely that the judge will have any time to read those reports until he considers Counsel's submissions at a pre-trial review, or actually starts sitting on the case.

The authors suggest that a Special Referee, who could be an expert or arbitrator appointed by the trial judge, could be asked by the judge to review the expert evidence and report to the judge any particular problems that may be apparent from reading the reports; e.g. where the experts have not addressed the same question. It cannot be in the public interest to have the inevitable waste of time which occurs during the course of a hearing when one side's expert is requested to address questions already addressed by the other side. The Special Referee could be given some jurisdiction and power in this area to ensure that the defaulting expert addressed specific questions, with the sanction that if the expert did not do so, his evidence would be struck out in part, or altogether if necessary.

11 The trial: the expert in the witness box

'The true fashion of the courts is . . . not to conciliate or exhort the parties, much less to hurry them . . . but to use the available machinery of litigation to enable them to settle their disputes according to law without grievous waste and unnecessary delay.'

Sir Francis Newbolt, Official Referee
(1920–1936)

This is the whole purpose of civil litigation procedure and arbitration. If the purpose fails then the worst that can happen is the adversarial battle of argument and evidence at a trial or hearing which proceeds to judgment or award. In that situation the expert must be careful not to take up confrontational attitudes.

The hearings in most, if not all, construction cases these days take place before Official Referees in the High Court and in arbitrations before construction arbitrators. Sometimes the Official Referees will sit outside London when circumstances demand. Arbitrators will, of course, sit at the convenience of the parties. In recent years there has been an increase in Official Referees' actions and in construction arbitrations. Some innovative practices and procedures have been introduced in order to shorten the waiting lists for hearing and to expedite matters.

It is now possible for an Official Referee to order the exchange of proofs of evidence. This practice has in fact proved to be of great assistance to the courts in shortening the time of hearings and particularly in reducing the length of time an expert witness or others spend in giving evidence in the witness box. Some of these innovations have already been implemented by a number of construction arbitrators.

Preliminaries

If the action or arbitration is a major one involving important points of law and substantial commercial risks, leading counsel may be instructed by the client's solicitors. In such cases the expert must be prepared to carry out various additional tasks to those outlined in previous chapters, although the expert will be aware of additional duties that may result from discovery. The expert must bear in mind that as important as his role may be, that of counsel presenting the case is critical. He is the leader of the case and everyone looks to him for a view as to how the matter will be generally conducted. He is responsible for the conduct of the case in court and together with the solicitor is responsible for presentation of the evidence. The expert's duty in this respect is to ensure that the opinion he gives counsel and any technical advice is balanced and objective. Any weaknesses in the technical case must be disclosed as has been emphasized previously in this book, and the expert must always be frank with counsel.

Whilst the expert is the expert in technical/scientific matters, the expert must accept that counsel is the expert in presentation of the evidence and how that presentation is conducted may vary from advocate to advocate. It would be impossible to describe the particular needs and requirements of counsel as these vary according to case and advocate.

Procedure at trial

Procedure has already been dealt with in Chapter 6. So far as form is concerned, as a matter of practicality an Official Referee may direct that both experts be sworn and each deal with the scheduled items in order, taking evidence of one, then of the other.

The court has power to direct that expert witnesses be called after all other evidence has been heard (*Briscoe* v. *Briscoe* (1966); *Barnes* v. *BPC (Business Forms)* (1975) considered; *Bayerische Ruckversicherung*

Aktiengesellschaft v. *Clarkson Puckle Overseas The Times*, 23 January 1989; *Alpina Zurich Insurance Co.* v. *Bain Clarkson* (1989).

Examination-in-chief

After leading counsel has opened the case, either by full speech referring to relevant evidence in chronological order, or by a speech submitted in writing with additional oral submissions (if agreed), the expert witness for the plaintiff/claimant is called after the witnesses of fact. He will take the oath in the prescribed manner, give his personal particulars, qualifications and experience and then deal with the technical issues of the case. Ideally this should be a spontaneous conversation between the expert and the leader, although the expert cannot lead the discussion.

The expert should be careful to answer the precise question and nothing else. He should make his answer clear and unambiguous, and narrate events in chronoglogical sequence. The rules of evidence outlined in Chapter 7 must be respected and every fact proved or opinion given must be relevant to the issues in the case. The expert will have with him for the purposes of such examination his final report or any other reports he may choose or counsel may advise. Sometimes evidence in chief is just that, because the expert is merely asked to produce his report. At the hearing the expert will usually be permitted to explain his opinion with particular reference to the real evidence, for example to samples in court or to drawings and sketches. A video recording has been introduced in evidence for the court before now but there is a limit to what one can substitute for oral evidence, bearing in mind that the judge or arbitrator wants to hear it from the man himself, not from mechanical or computer devices.

In examination-in-chief counsel is not permitted to ask leading questions: i.e. questions that suggest the answer, except to identify the witness and to give his experience and qualifications or to refer to facts which have to be proved. The expert may be asked a leading question in examination-in-chief if, for example, the other side's expert alleges he said: 'I cannot find any damage as a result of this leakage.' The expert can then be asked in examination-in-chief: 'Did you ever say to Mr X, the defendant's expert "I cannot find any damage as a result of this leakage"?'

Counsel should ask short questions which may enable the expert to give a full explanation of his investigations and tests carried out and the reasons for his opinions. However, the expert should bear in mind that, when he is giving his evidence in this way, the defence or opposing counsel will be noting what he says: any contradiction with his report or with his evidence will be challenged in cross-examination.

Cross-examination

Cross-examination can be a daunting task for those who do not know their case. It should hold no terrors for a thoroughly prepared expert. However, he must prepare himself as far as possible for the unexpected. Sometimes the unexpected question can be dealt with simply and effectively. A great deal depends on the expert's ability to master the facts and to deal with the technical issues.

The duty of cross-examining counsel is to deflect the strength of the expert's evidence adduced in examination-in-chief – to chip away at it, so to speak, so that the opposing expert can dissect it and destroy technical arguments and the foundations for the allegation. One often hears it said that the days of the great lawyers are past but the skills of the cross-examiner are as high as ever, and cross-examination by leading construction counsel should never be underestimated.

The most decisive weapons of the cross-examiner are:

- repetition of question
- ridicule of the witness, and
- the tactical manoeuvring of a witness so as to ensnare him in an admission favourable to the cross-examiner's client.

These weapons can be used in various guises with questions politely put but potentially lethal. The expert witness must be able to deal with this.

Counsel on the other side may go out of his way deliberately to confuse you with facts, figures, dates and assumptions. This is to be expected. The expert must deal with this. If the expert is proved wrong, or has good reason to correct himself or change his opinion, then he must do so otherwise he will lose his credibility and at worst his integrity.

The expert must never answer a question unless he fully understands it.

The expert may be asked for calculations while in the box. This may not be practically possible and if he is any difficulty he or counsel should make appropriate representation to the judge that the task is inappropriate and/or impracticable. Judges have a discretion in these matters.

Perhaps the most famous example of examination of an expert witness occurred in a murder trial when the great advocate Norman Birkett KC, as he then was, was prosecuting the criminal Rouse for murder (see *The Art of the Advocate* by Richard Du Cann). The prosecution's case hinged on the finding of a brass nut, discovered loose on a petrol pipe after a fire destroyed a car in which the victim had been placed by the murderer.

The defence called an expert to claim the fire had been started accidentally and that the nut came loose in the fire. Birkett challenged the expert by asking:

'What is the coefficient of expansion of brass?'

Witness: 'I am afraid I cannot answer that out of hand.'

Birkett: 'If you do not know, say so. What do I mean by the term?'

Witness: You want to know what is the expansion of the metal under heat.'

Birkett: 'I asked you what is the coefficient of the expansion of brass. Do you know what it means?'

Witness: 'Put that way I probably do not.'

Birkett: 'You are an engineer?'

Witness: 'I dare say I am.'

Birkett: 'Well you are not a doctor or a crime investigator or an amateur detective are you?'

Witness: 'No'.

Birkett: 'Are you an engineer?'

Witness: 'Yes.'

Birkett: 'What is the coefficient of the expansion of brass? Do you know?'

Witness: 'No, not put that way.'

The witness here cut a pathetic figure and not one whose opinions would be given much weight by the court.

Birkett used the weapons of ridicule and repetition to pin down his witness who, being an engineer, inevitably felt embarrassed at not being able to answer an apparently straightforward question.

The witness might have been forgiven for not knowing the exact calculation to six decimal places of the coefficient of the expansion of brass, but should at least have given the court the impression that he knew how to obtain the figure, even if could not remember it at the time. One of the great attributes of an effective expert witness is that he does justice to his client, the court and the standards of his profession. If he is not of sufficient standing in his profession, it is doubtful whether he will be of any standing as an expert in court.

Under cross-examination the expert cannot be compelled to disclose privileged information, advice or communications between himself, counsel and solicitors.

Re-examination

The purpose of re-examination is to enable the witness to clarify or correct any statement made during cross-examination that may have been given a misleading emphasis by cross-examining counsel. The judge may ask a few questions to clarify matters where he feels it necessary. It is not the usual practice of judges to 'enter the arena' of the court but recently some lawyers have advocated this somewhat inquisitorial attitude in the hope that trials could be shortened. It must, however, be a matter for the discretion of the judge and what he feels is necessary in the circumstances.

Summing up

At the close of the evidence counsel for the defence will sum up. This is followed by the closing speech for the plaintiff. The judge will then consider the evidence and may adjourn the proceedings to consider his judgment. The adjournment may take some weeks while the judge writes his judgment. The court will then re-convene at a fixed time and judgment will be given. In smaller cases, such as Order 14 matters, the judge may give judgment as soon as he has heard the case.

The expert should carefully review the judgment and see whether his views were accepted. It is interesting for the expert to compare the points made in his report with the points the judge finds proved and accepted. It is also useful for the expert to learn exactly how his views were interpreted and see what lessons can be learnt for the future.

Checklist for the trial

   (1) Remember your duty is to give an honest, clear and concise view of the matter.

   (2) You must prepare yourself thoroughly for giving evidence and be prepared to answer all technical questions related directly or indirectly to the matters and issues.

   (3) Part of your duty will be to assist leading and junior counsel at the trial and to be present when the other side's expert is giving evidence. If there are negotiations on a without prejudice basis outside the court room, your presence may be required.

   (4) Once you have taken the oath, you must not speak to anyone, including counsel, solicitor or client, about the case. Once you have finished your evidence you may resume your role of assisting the legal team.

   (5) In the witness box only answer what counsel asks. Do not volunteer additional information unless clearly invited.

   (6) If you have any doubts about procedural difficulties or requirements of proof all such questions should be thoroughly discussed beforehand with counsel.

   (7) You are not expected to read your report line for line in the box. You may do so before an arbitrator if the other side agrees, or it may be taken as read by the Official Referee. You must, however, explain every allegation you have made, the course of your investigations and your findings. You may refer to research material, professional journals, codes of practice,

Building Regulations, textbooks or statutory authorities to support your views.

(8) When giving evidence at the hearing you should address the judge or the arbitrator directly, not counsel.

(9) In Official Referees' cases and before arbitrators it is important that you speak clearly at a pace which allows the judge or arbitrator to make a full note of what you say. Watch the judge's pen.

(10) Do not be overawed by cross-examination. Deal with each question as it arises and give a clear, precise and direct answer.

(11) If there is anything you have said which you subsequently consider wrong, whether in your report or your oral evidence, in chief or in early cross-examination, it is your duty to say so and correct any such matter you consider to have been in error.

12 'Che sera sera'

General

This book has generally outlined the law relating to expert evidence and the practice that may be adopted by experts as witnesses. It is the role of the expert as a witness of fact and pricipally one of opinion that has been addressed. But what of the future? What will be the effects of EC legislation? Will there be a trend towards strict liability on an ever increasing scale? Will the law relating to warranties assume a greater significance than the law of negligence? These are not so much matters for experts as for lawyers, but experts will inevitably play their part in their determination.

As experts look towards the future, with the knowledge gained from their various experiences before various tribunals, they must be wary of any 'quick-fix' solutions which do not do justice to their clients, nor to themselves or their professions. With the introduction of more EC legislation matters may become more complex. The expert must be conscious of EC implications, and this book cannot ignore EC legislation or its implications for experts. Neither can this guide for experts ignore the use of experts in other legal systems such as those in France and Germany.

The expert and the EC dimension

The European Community has as its chief aspiration the harmonization of commerce within the community. Within that general overriding aspiration is a need to harmonize the legal duties and obligations under the laws of member states. Those are ideals first pursued in Roman times, resurrected in the Napoleonic era, and now being reflected in some Member States' legislation. The concept has been described as a concoction of 'Greek philosophy, Gallic logic and English pragmatism'. If the expert witness appreciates those aspects then he may be able to appreciate the difficulties of interpeting EC directives and reconciling Member States' laws. Of direct interest to experts is the product liability directive and EC proposals for a directive on the liability of those who supply services for others.

Product liability

The directive of the EC on product liability was implemented in the United Kingdom by Part 1 of the Consumer Protection Act 1987. Liability is strict, i.e. there is no defence on the basis that all reasonable care was taken. It does not in any way replace existing common law liability or negligence but gives additonal rights to persons aggrieved.

Establishing liability

Liability is established by proof of a defect in the product. A 'product' includes goods which may be substances (natural or artificial), crops, fixtures to land, ships, vehicles, and aircraft. A 'product' also includes electricity and a product which may be comprised in another product, whether it is raw material or otherwise.

Persons affected

Persons who may be liable are: producers, persons holding themselves out as producers by marking a product with a trademark or other mark, and those who import products into Member States, section 2(2) of the Consumer Protection Act 1987.

Whether a contractor can be a producer must be a matter of interpretation of section 1(2) of the Act. This defines a 'producer' in relation to a product as:

(1) the person who manufactured it;
(2) in the case of a substance which has not been manufactured but

has been won or abstracted, the person who won or abstracted it;

(3) in the case of a product which has not been manufactured, won or abstracted but essential characteristics of which are attributable to an industrial or other process having been carried out (e.g. in relation to agricultural produce), the person who carried out that process.

The EC directive may be helpful in understanding the definition's scope insofar as Article 3 defines a 'producer' as 'the manufacturer of a finished product . . . or the manufacturer of a component part', so that a contractor may come within the definition.

Supplying goods is defined in section 46(1) of the Act as being *inter alia*: '(c)the performance of any contract for work and materials to furnish the goods'. This also appears to include a contractor. This should be read in the context of section 46(3) which provides that the performance of any contract in the erection of any building or structure on any land or carrying out of any other building works is treated for the purposes of the Act as 'a supply of goods' only insofar as it involves the provision of any goods to any person by means of their incorporation into the building structure or works. This means in effect that a contractor may be liable both as supplier and producer.

The expert's function

As experts might expect, the 'state of the art' defence applies. Section 4(1)(e) of the 1987 Act provides that it shall be a defence:

'that the state of scientific and technical knowledge at the relevant time was not such that a producer of products of the same description as the product in question might be expected to have discovered the defect if it had existed in his products while they were under his control . . .'

In other words there may be plenty of work for those experts who may be asked whether a defective product ought to have been discovered.

Liability for services

The draft EC Directive on liability of those who supply services for others should become law in the UK by 31 December 1992. This proposal, and it must be emphasized at the date of writing that this is still a proposal, may introduce a revolutionary shift in the burden of proof from the plaintiff to the defendant in an action. That is to say it will be for the defendant to show that he was not at fault in supplying a service.

Article 1 of the EC proposal means that the supplier of the services will be liable for damage caused whilst supplying that service unless he can prove that he was not at fault in supplying that service. The Article further provides that in assessing the fault, account must be taken of the behaviour of the supplier of the service and the degree of safety one ought reasonably to expect from him.

Article 2 defines a 'service' as:

'any transaction carried out on a commercial basis or by way of a public service and in an independent manner, whether or not in return for payment, which does not have as its direct and exclusive object the manufacture of movable property or the transfer of rights in rem or intellectual property rights.'

Three areas of liability are excluded from the definition namely:

(1) public services intended to maintain public safety;
(2) package holidays; and
(3) services concerned with waste.

The proposed directive concerns physical protection of persons and property, and not the person's economic protection, so that such matters as investment advice, and surveys of property would be by definition excluded from the scope of the directive.

A 'supplier' is defined by Article 3 as 'the natural or legal person who provides a service in the course of his commercial activities or public functions'. If such functions are sub-contracted in whole or in part, then the independent sub-contractor will be considered as a supplier of services and will be liable for damage caused by his fault.

Damage

Article 5 provides that the injured person is required to establish that the performance of the service caused him damage. Article 4 provides that damage will mean:

(1) death or any damage to the health or physical integrity of persons;
(2) any damage to the physical integrity of their movable or immovable property, including animals; and
(3) any financial damage resulting from the damage referred to in (1) or (2) above.

Significantly the proposed directive does not included pure economic loss such as loss of profit within the definition of damage. The definition extends only to private property, it does not cover commercial property.

A supplier of a service is liable for damage to the physical integrity of immovable property, including the property which was the object of the service (Article 1). If a civil engineer therefore was responsible for the design of underpinning a private dwellinghouse which subsequently failed, then the building service performed by the engineer would presumably fall within the scope of professional services defined within Article 1. If the client proceeded to bring an action against the civil engineer, under enabling legislation implementing the directive along the lines of the proposal, the engineer would have to prove that his underpinning design specification etc. was not at fault.

A supplier of services may not limit or exclude his liability (Article 7). This is a similar provision to section 2(1) of the Unfair Contract Terms Act 1977.

Article 8 provides that all persons responsible for specific damage are jointly and severally liable.

Under Articles 9 and 10 there is a limitation period of three years from the date upon which the plaintiff became aware, or ought reasonably to have known of the damge. If no action is commenced within five years the right of action is extinguished. In cases of design and construction of buildings these periods are extended to ten and twenty years.

The above analysis is based upon the latest draft of the proposed directive. The reader should be aware that the draft is subject to revision and could quite possibly shift in emphasis from the narrow consumer protection view to the wider view of liability of suppliers of services.

Implications for expert witness

The victims of defective services will often find it difficult to prove that the supplier of the service is at fault because the supplier, having the technical knowledge, can usually provide proof to the contrary far more easily. This should mean that there will be an increase in demand for experts to give their opinions as to the quality of such services supplied. Whilst liability for defective products is now strict, subject to certain defences following the implementation of EC Directive 85/374/EEC, the proposal for service liability will be fault based. Whereas in the latter case the expert will be required to give his opinion on whether the particular duty was observed or not, in the former case the expert will be retained to state whether in his opinion his client may fall within one of the specific categories liable for the defect.

It is impossible to predict the precise course that English courts may adopt in their interpretation of these highly complex matters, but apparently simple questions such as the meaning of 'damage' and 'damage to physical integrity' may produce some intriguing and possibly complicated debate. The expert must be careful in interpreting directives and seek legal opinion if in any doubt. It is noted, for example, that the word 'regulation' has different meanings in the Member States. There may well be a marked increase in cases concerning 'fitness for purpose' or 'intended use', and whether a product has met an appropriate EC standard. What would be suitable as a standard e.g. for a tile manufactured in Southern Spain may not be of particular benefit in guaranteeing its use in e.g. Cleveland.

The European Court

The implementation of EC directives also opens up a Pandora's box of a possible right to appeal to the European Court under Article 177 of the treaty establishing the European Economic Community; Article 150 of the treaty establishing the European Atomic Energy Community, or Article 41 of the treaty establishing the European Coal and Steel Community. Order 114 of the Rules of the Supreme Court provides that the court will only give leave for appeal in exceptional circumstances and these cases must be very rare indeed. Nevertheless experts ought to be aware of the right bearing in mind the difficulties that may arise in interpretation. The expert's opinion may not be the end of the matter.

An order of reference under the EEC and Euratom Treaties will be made by the court where the following two conditions are satisfied:

(1) where a question concerning the validity or interpretation arising under these treaties 'is raised before the court' (see Article 177 of the EEC Treaty). Such a question may be raised by either party or by the court itself. It will usually be raised in the pleadings but it may be raised in some other way e.g. in the affidavits of the parties; and

(2) where the court considers that 'the decision on the question is necessary to enable it to give judgment' (see Article 177 of the EEC Treaty).

Article 177 of the EEC Treaty provides as follows:

'(1) The Court of Justice shall have jurisdiction to give preliminary rulings concerning:

(a) the interpretation of this Treaty;
(b) the validity and interpretation of acts of the institutions of the Community;
(c) the interpretation of the statute of bodies established by an act of the Council where those statutes so provide.

(2) Where such a question is raised before any court or tribunal of a Member State, that court or tribunal, may if it considers that a decision on the question is necessary to enable it to give judgment, request the Court of Justice to give a ruling thereon.'

If an expert finds that a matter has been appealed to the European Court, his correspondence with the lawyers may not be privileged as it is in English law, and it is for the authorised agents of the European Commission to determine whether any claim for privilege in that respect might be justified.

Professional qualifications

An EC directive of 21 December 1988 which came into force on 4 January 1991 enables non-UK nationals who possess the appropriate professional qualifications and training to join an appropriate UK professional body. This directive does not of course simply enable non-UK nationals to join UK professional bodies but also permits UK professionals to join European institutions.

Quo vadis peritiae?

Those who may aspire to a wider role may be interested in considering the role of the expert witness in France, Germany and Scotland. It is also interesting to compare the role and function of the expert in France, Germany and Scotland with that of the traditional role of the expert in England, in particular in relation to the function of the court expert where some lessons can be learned.

The expert witness in Scotland

The first edition of this book did not address directly or indirectly the role of the expert witness in the Scottish courts. Chapter 4 of this edition deals with the judgment in *Davie* v. *Magistrates of Edinburgh* (1953). This case is accepted as an authority on the role of an expert witness and is viewed by the authors as one of the most perceptive judgments dealing with that important point.

Like the English expert witness, an expert witness appearing in the Scottish courts is only required to assess certain facts in their specialist context and explain those as such to the court.

The Scottish courts accept that, as in England, the courts have need for more expertise in technical matters where science is advancing faster by the decade than in previous centuries. It is, however, felt to be important to ensure that the traditional judicial process is not simply replaced by a trial by experts. As Lord President Cooper explained in *Davie* v. *Magistrates of Edinburgh* (1953):

'the parties have invoked the decision of a judicial tribunal, and not an oracular pronouncement by an expert.'

The Scottish cases possibly have a parallel to *R* v. *Silverlock* (1894) in *Hopes and Lavery* v. *HMA* (1960) where Lord Sorn held that a typist might have become an expert on the authenticity of a transcript from a tape recording on which she had worked, because there was no rigid rule that only witnesses possessing some technical qualification could be allowed to expound their understanding of any particular item of evidence.

In Scotland there is no clear authority that an expert witness cannot give an opinion on the main issue before the court, and it is considered that this may usurp the function of the trial judge. However in *Morrison* v. *Maclean's Trustees* (1862) particular witnesses were asked to comment upon the mental state of the testator at the time he signed his will. No objection was taken to the examination of the executors by counsel as to medical evidence on the testator's mental state, but the trial judge directed the jury that the issue of capacity to sign was a matter for the jury and not for the medical witnesses. The jury could therefore disregard the opinions of the medical experts and substitute their own judgment to that question.

Like English courts, Scottish courts are frequently asked to interpret terms of a contract and in such matters they apply the ordinary law whereby words must be given their normal and natural meaning as used by the parties without the aid of any expert witness (*Tancred Arral and Co.* v. *Steel Co. of Scotland* (1887)).

Possibly one of the greatest revolutions in Scottish civil evidence has been the total abolition of the rule requiring corroboration of evidence in civil proceedings, insofar as the rule had not been abolished by previous legislation. Section 1 of the Civil Evidence (Scotland) Act 1988 now provides such a far reaching change. A Scottish judge, Lord Wark, said: 'the testimony of one witness, however credible, is not foolproof of any ground of action or defence, either in a civil or a criminal cause'. But the rules of evidence have indeed advanced a considerable way since then.

Other amendments to the Scottish law of evidence made by the Civil Evidence (Scotland) Act 1988 provide for the admission of hearsay evidence in civil proceedings, and authorize courts hearing civil actions to conclude that a fact has been proved even though the only evidence of the fact comes from hearsay evidence (section 2). Previous statements made by a witness in civil proceedings (except those contained in precognitions) are admissible for the purpose of either attacking or supporting a witness's credibility (section 3). Courts are enabled to recall a witness who has already given evidence, or call a fresh witness and hear evidence for the purposes of sections 2 and 3 of the Act (section 4). Provision is also made for suitably authenticated documents to be regarded as part of the records of a business or undertaking (section 5). The admission of evidence of suitably authenticated copy documents is permitted to the same extent as the original (section 6), and provision is made for the admission of evidence to the effect that a particular document has never been part of the records of a business or undertaking, without the need to produce those records (section 7).

The court expert in France

In France, as a matter of practice, the parties are free to choose their own expert. In certain cases, however, the court may appoint an expert. The court expert is the 'eye of the judge' ('l'oeil du juge'). The expert is appointed by the court from a list of experts for the particular court. The Court of First Instance, and the Court of Appeal appoint the expert from the court list of experts drawn up by the Court of Appeal and the Cour de Cassation (the regional and national list).

The court expert must act with 'conscience, objectivity and impartiality'. He must not be engaged in any conflicting activity. He cannot be an advocate, a judge or a bailiff. He does not have to have French nationality, he may be an EC national, or a non-EC national subject to international convention.

The court expert swears:

'to give the contribution to justice, to fulfil his mission, to write his report and to give his opinion to the best of his honour and conscience'.

The fees of the court expert are payable by the party bringing the case. Fee levels are determined by the type of expertise required and the function required. The fee is payable once the expert has completed his report and may be subject to taxation by a taxing judge. The costs claimed must not be unreasonable and will depend upon the type of case or matter and the extent of the expert's involvement.

The expert is liable for misconduct and subject to discipline by the court. The expert can be struck off the court list or can be temporarily suspended from practice as an expert.

Not all expert witnesses are appointed by the judge. Parties can choose an 'amiable' expert or a 'non-official' expert but the court appoints a 'judicial' expert. The latter is always the servant of the court.

The court expert in Germany

In the German courts it is the judge who appoints the expert witness. The judge will brief the expert who will produce a report for the court. This will be put into evidence. It is the judge, however, who examines the expert in respect of his evidence.

When the expert gives his evidence he is told and cautioned by the judge that anything he says will be taken down in evidence and he can be liable for what he says. He is cautioned to be 'as impartial, fair and diligent' as he possibly can. This is slightly different from the English system where the expert is simply sworn as a witness and told to tell the truth the whole truth and nothing but the truth.

When appointing a court expert, the judge will often go to the local Chamber of Commerce (Industrie und Handelskammer) for names of appropriate experts. One side may object to the court expert and if the objection is sustained, the judge may appoint what is termed a counter-expert.

The rules as to expert evidence and its admissibility are governed by the Code of Civil Procedure.

The court expert in England

There is perennial debate on the time and cost of litigation and arbitration, and an increasing number of suggestions for remedy made by contributors to 'Arbitration' the journal of the Chartered Institute of Arbitrators. Judges themselves have frequently over the last few years given public hints of experimentation with various ideas such as possibly adopting inquisitorial techniques. The adoption of inquisitorial techniques by judges would be extremely difficult in a system which is adversarial. There is some evidence that the Official Referee's court, a court that is of its nature highly sensitive to implications of cost and time, bearing in mind the highly complex technical nature of the voluminous evidence which inevitably comprises a case, is giving increasing encouragement to such matters.

The Official Referees have made many brave attempts to cut down wasted judicial time by various devices, such as exchange of experts' reports and ordering experts to make a statement as to what is agreed and what is not agreed as matters of expert evidence. They have also endeavoured to manage cases by dividing them where necessary into particular issues. All these brave efforts, however, are seen in some quarters as not being enough and more is demanded by the construction industry. It is inevitable that the Official Referees will look at the users of their service and look to them for some ideas. Possibly the use of Order 40 of the Rules of the Supreme Court in appointing a court expert would save further time and costs.

In a recent speech, Lord Murray stated that the courts in Ulster had been making orders for the appointment of a court expert since 1984. Judging by what Lord Murray said of his experiences it may well be that further benefit could be derived if English courts followed the Irish example.

It is suggested that as a matter of practice in certain cases (especially straightforward ones) standard directions could be issued to the effect that both parties agree on the judge appointing a court expert agreed by the parties, or that the judge should exercise his power under Order 40 of the Rules of the Supreme Court and appoint an expert in default of agreement, after consultation with the president of the appropriate institution. The institution would be the appropriate professional body, i.e. RIBA, ICE, RICS etc. The institutions have lists of experts who have had court experience. The British Academy of Experts also maintains a list of experts covering a wide variety of disciplines.

It is understood that the practice of county court district judges in family cases where the valuation of property is in dispute, is that the judge may appoint a valuer after consultation with the appropriate professional institution. It is suggested that Official Referees might adopt a similar system whereby their clerks or a registrar would consult the president of the appropriate institution and obtain an appropriate list of names so that the judge could choose a name from that list.

How a list of experts in each institution is compiled must of course be a matter for the professional institution concerned. The institutions regulate their members and are responsible for maintaining professional standards. What happens in practice is that members may notify their institution that they would like to give their names as experts and they are listed accordingly. The institution may then give a list of some names to the registrar or clerk of the court with details of that expert's experience as disclosed by the expert. The registrar could then suggest the names to the parties concerned and if agreed the expert might be appointed.

If there was any dispute as to the appointment of an expert, it might be useful if the clerk or the registrar (who should be a solicitor of some years practice) would then choose the expert or require a further list of names. These are technicalities and matters for obvious discussion and debate between the users of the court and the professional institutions. Current developments dictate that there should be open discussion of this issue.

The important point from the institution's point of view is to emphasize that the member of the institution holds himself out as an expert, but that the institution concerned is not vouching that that member is an expert. Indeed the institution could not do so because there cannot be any classification of expertise for the purposes of the law of evidence in England. See the section below dealing with 'a profession of experts'. It is simply a matter for the judge or the arbitrator to admit the evidence and prefer one expert's evidence to the other or, in the case of a court expert, to accept or reject it. See also the comments made in Chapter 1 when the question as to how an expert became an expert witness was considered.

Some guidance may also be obtained from Scotland where judges can appoint an 'assessor' or 'a man of skill' to give an expert report. The parties have a right to apply to the court by way of motion requiring that a named expert be appointed or agreed. If there is no agreement between the parties then the judge can appoint. It is understood that once the judge has received the report, he can adopt it and it will become part of his judgment in the case.

Appointment of court experts in the county court – form of order

It is understood that it was the practice in the Deptford County Court to appoint referees or assessors for enquiry and report in various cases. The form of precedent shown overleaf was used.

Present Rules

Under Order 40 of the Rules of the Supreme Court the court may at any time on the application of any party appoint an independent expert. If more than one such question arises as to expert evidence, additional experts may be appointed who are retained to enquire and report into any questions of fact or opinion which do not involve matters of law or questions of construction of documents. If the parties do not agree on the court appointed expert, then the court nominates the expert.

The expert is bound to send a copy of his report to the court (Order 40 Rule 2) together with the requisite number of copies as the judge may direct. The court must then send copies of the report to the parties' solicitors.

Under Order 40 Rule 3 if the court expert requires experiments or tests, then he informs the solicitors concerned and advises them as to the amount of expenses. If the parties cannot agree then the matter must be decided by the court.

Any party has the right under Order 40 Rule 4 to apply to the court for leave to cross-examine the court expert on any matter contained in his report.

Under Order 40 Rule 5 remuneration of a court expert is fixed by the court and will include a fee for his report and a sum for each day of his

Form of precedent

IN THE   COUNTY COURT   Plaint No:

B E T W E E N:-

....   **Plaintiff**

and

....   **Defendant**

TO ALL PARTIES:

IT IS ORDERED, pursuant to Section 65 of the County Courts Act 1984, and Order 19 of the County Court Rules 1981 that these proceedings and all questions arising therein be referred to Mr.   of
           for enquiry and report.

AND IT IS FURTHER ORDERED THAT:

(1) All the pleadings herein be sent to Mr.   (hereinafter referred to as the Referee) forthwith by the court.
(2) The court do supply the Referee with the terms of section 65 of the County Courts Act 1984 and of Order 19 of the County Court Rules 1981.
(3) The Referee shall conduct his inquiry in such manner as he considers to be fair and just.
(4) The Referee shall submit his report to the court in writing by not later than the   day of   1992.
(5) The Referee may submit any question arising in the course of his inquiry for the decision of the court.
(6) The report of the Referee shall deal with the following matters:

(a)
(b)
(c)
(d)

AND IT IS ORDERED that these proceedings stand adjourned for the consideration of the report until the   day of   1992.

DATED the   day of   1992

Judge

attendance. The parties, however, remain jointly and severally liable to pay the amount fixed by the court for the expert's remuneration. If the appointment of the court expert is opposed then the court may require the party requiring such appointment to give security for the remunertion of the expert as the court thinks fit.

It was Lord Denning in *Re Saxton* (1962) who expressed the hope that in future careful consideration might be given to the appointment of a court expert.

Litigants, as a matter of principle, must be free to choose their own expert unless it is a particular case where both commerical and legal interests would dictate that a court appointed expert would provide the most appropriate means of achieving justice in the case.

The *Guide to Commercial Court Practice* (1986) states (at para 15.6) that:

> 'in cases with a high scientific content consideration should be given to saving time and cost by means of assessors (Order 33, Rule 2(c)) or the appointment of a court expert (Order 40).'

The chief problem at the moment with Order 40 is that there are no totally effective cost provisions. When an Official Referee recently appointed a court expert he had a particular difficulty over costs. He ordered payment of the expert's fees out of monies already paid into court. If the monies, however, had not been paid in, the judge would have experienced some embarrassment if he had appointed the expert and the party against whom the order was made could not afford to pay the expert. It is understood that the Treasury will not appropriate any funds for such purpose and consequently the burden of costs must fall upon the parties. Anyone appointed as court expert should, therefore, not presume that his costs will be paid at the end of the day and ought to be put on notice that his costs may never be paid at all.

A profession of experts?

The fundamental principle expounded by Lord Chief Justice Mansfield in the 18th century was: 'The opinion of scientific men upon proven facts may be given by men of science *within their own science*'. It is clear from that precise definition that those few words have more importance and significance than all the volumes of text books on expert evidence. Thus, the expert must give evidence within his own area of expertise, i.e. from his own professional discipline, governed by his own profession's standards, his profession's code of practice and discipline and no more.

It goes without saying that if e.g. an architect gives expert evidence he gives it as an architect and tells the court as simply as he can what principles of architecture will apply, why they apply and how they apply to the particular problems in issue in the action. The expert architect does not need any qualification as an expert *per se*. He has the qualification by reason of his own professional discipline. It therefore follows that persons giving expert evidence do not require any qualification as members or fellows of any association of experts in order to be competent expert witnesses. Whether the witness is an expert, or 'man of skill' as they say in Scotland, is entirely a matter for the court or the arbitrator. It is not a matter for any other body, however well intentioned, save the expert's own professional body, e.g. the RIBA.

For many years the Chartered Institute of Arbitrators (which was founded by two eminent solicitors during the First World War) has run many excellent courses on expert evidence. In the construction industry many professionals qualify as associates or fellows of the Chartered Institute of Arbitrators to develop their expertise in a judicial capacity as arbitrators. They are generally termed 'experts', but in practice one finds that they do not like the term 'experts' and prefer to be described according to their professional discipline. The use of the term consultant rather than 'expert' is preferred in England.

Over recent years there has been debate about the expert's role in the Chartered Institute of Arbitrators. Whilst seminars, talks, discussions and mock trials are useful, those who would give evidence in court as experts must appreciate that there are no substitutes for actual court experience. See the comments made in Chapter 1.

The client inevitably turns to the established professional institutions for expert opinion. Firms of solicitors generally have their own lists of experts and if there is no appropriate expert on the list they will consult their colleagues or counsel and otherwise, possibly, the Law Society or other professional body governing the appropriate particular profession, e.g. RIBA.

In conclusion it must be said that it is difficult if not impossible in law to specify classes of expertise and lay down any specific guidelines for various professional institutions. That is inappropriate and unwise. The best that can be said is to reiterate the conclusion reached by the Law Reform Committee in its 17th report dated October 1970 (Pearson report) which recommended that:

'it was undesirable to lay down as a *matter of law* any minimum qualification in any class of expertise.'

To go beyond or outside that criterion would not be within the scope of this book on expert witnesses.

'Peritia'

The adversarial system which has evolved in England over the centuries, and the common law which has been built with care and precision but which has remained adaptable and flexible has imbibed many continental influences. It is recorded in the annals of Pollock and Maitland that after the Romans left England various continental influences from Iceland to Spain influenced formation of local customs and laws. A system which has survived all those influences over the centuries will certainly survive in its adversarial form any influences from Europe or the United States.

Within that vast system expert witnesses have a role to play. It is an essential role but a limited one as expert witnesses can never be a substitute for lawyers, they cannot act as advocates, nor are they competent to act as judges. They do, however, have an overriding essential use in translating technical complexity to simple form for the benefit of the court.

A good expert must remain a person who is good professionally, who has a general grasp of the principles of the law of evidence but is not a lawyer, recognizes court procedures and disciplines, and who if called upon to give his opinion can give his opinion openly, honestly and fully to the court or arbitrator. Above all, as Lord Russell of Killowen said, in *R* v. *Silverlock* (1894), the expert witness must be *'peritus'*, i.e. skilled.

Glossary

ACTION

A proceeding by which one party seeks to enforce some right against another party, or to restrain the commission of some wrong by another party against that party. It is usually commenced by writ or a form prescribed by the rules of court.

AFFIDAVIT

A written statement of fact affirmed or sworn by a deponent on matters within his knowledge or belief.

ARBITRATION

The determination of a dispute or difference between two or more parties by an agreed or nominated arbitrator.

ARBITRATOR

A person or persons appointed by agreement between the parties to a dispute to resolve matters in issue between them.

AWARD

A final judgment upon which all matters referred to the arbitrator are determined unless the agreement for arbitration expressly provides otherwise.

CLAIMANT

The party making a claim.

DEFENDANT
A party who defends the action.

DIRECTIONS
Orders given by the judge, Official Referee or arbitrator for the conduct of the proceedings.

DISCOVERY
A process by which the parties to an action or arbitration are required to disclose all material documents in their possession.

INJUNCTION
An equitable remedy whereby a person is ordered to refrain from doing, or is ordered to do, a particular act or thing.

INTERLOCUTORY
Proceedings which take place between the service of a writ and the trial, or between the service of notice to concur in the appointment of an arbitrator and the hearing.

LIQUIDATED AND ASCERTAINED DAMAGES
A fixed amount of damages agreed by the parties to a contract prior to the execution thereof.

OFFICIAL REFEREES
Circuit judges who principally hear construction cases but who also have jurisdiction to hear matters of technical or scientific complexity.

PAYMENTS INTO COURT
Payments into court are usually paid in by a defendant who has no defence or limited defence to an action. It is a tactical process whereby the plaintiff may be put at risk as to costs if he continues with the action and does not obtain a judgment in excess of the amount of the payment in.

PLEADINGS

Written statements of fact containing the respective parties' cases. They comprise such documents as statement of claim, defence, counter claim, defence to counterclaim, and reply.

POINTS OF CLAIM

The claimant's pleading which sets out the facts upon which he will rely to prove his case in an arbitration.

PRIVILEGE

A right not to disclose certain classes of documents or communications, e.g. legal advice.

RESPONDENT

The person who defends the arbitration.

SEALED OFFER

The equivalent of a payment into court in an arbitration. The practice of sending the arbitrator a sealed envelope containing an offer. The arbitrator is directed not to open the sealed envelope until he has made his award.

STATEMENT OF CLAIM

The plaintiff's pleading which sets out the facts in issue upon which he intends to rely at the trial.

TAXATION OF COSTS

If the unsuccessful party does not agree the successful party's costs, he may require the successful party to have his bill of costs taxed. In litigation this is carried out by a taxing master of the High Court, and in arbitration either by a taxing arbitrator applying the rules of the Chartered Institute of Arbitrators, or by a taxing master of the High Court if the other party so requires.

THIRD PARTY

A person who is joined in the proceedings by the defendant where there are issues of legal liability as between those parties.

Key to law reports

AC	Law reports, appeal cases, House of Lords
All ER	All England Law Reports
ALR	Argus Law Reports
BLR	Building Law Reports
Ch	Law Reports, Chancery Division
Ch App	Law Reports, Chancery Appeals
Ch D	Law Reports, Chancery Division
CILL	Construction Industry Law Letters
CLR	Common Law Reports
CLR	Commonwealth Law Reports
Con LR	Construction Law Report
CONST LJ	Construction Law Journal
CL&F	Clark and Finnelly
EG	Estates Gazette
Exch	Exchequer Reports
H&N	Hurlstone and Normans Reports, Exchequer
HL	House of Lords
JC	Justiciary Cases
Lloyds Rep	Lloyds List Law Report
QB	Queens Bench Reports
QBD	Law Reports, Queens Bench Division
RPC	Reports of Patent Cases
SOL JO	Solicitors Journal
SC(HL)	Court of Sessions Cases (Scotland) (House of Lords)
SLT	Scotts Law Times
WLR	Weekly Law Reports

Order 40
Court Expert

Appointment of expert to report on certain questions (O.40, r.1)

1.—(1) In any cause or matter which is to be tried without a jury **40/1**
and in which any questions for an expert witness arises the Court
may at any time, on the application of any party, appoint an
independent expert or, if more than one such question arises, two or
more such experts, to inquire and report upon any question of fact
or opinion not involving questions of law or of construction.

An expert appointed under this paragraph is referred to in this
Order as a 'court expert.'

(2) Any Court expert in a cause or matter shall, if possible, be a
person agreed between the parties and, failing agreement, shall be
nominated by the Court.

(3) The question to be submitted to the court expert and the
instructions (if any) given to him shall, failing agreement between
the parties, be settled by the Court.

(4) In this rule 'expert,' in relation to any question arising in a
cause or matter, means any person who has such knowledge or
experience of or in connection with that question that his opinion
on it would be admissible in evidence.

Report of Court expert (O.40, r.2)

2.—(1) The Court expert must send his report to the Court, **40/2**
together with such number of copies thereof as the Court may
direct, and the proper officer must send copies of the report to the
parties or their solicitors.

(2) The Court may direct the Court expert to make a further or
supplemental report.

(3) Any part of a Court expert's report which is not accepted by all the parties to the cause or matter in which it is made shall be treated as information furnished to the Court and be given such weight as the Court thinks fit.

Experiments and tests (O.40, r.3)

40/3   3. If the Court expert is of opinion that an experiment or test of any kind (other than one of a trifling character) is necessary to enable him to make a satisfactory report he shall inform the parties or their solicitors and shall, if possible, make an arrangement with them as to the expenses involved, the persons to attend and other relevant matters; and if the parties are unable to agree on any of those matters it shall be settled by the Court.

Cross-examination of Court expert (O.40, r.4)

40/4   4. Any party may, within 14 days after receiving a copy of the Court expert's report, apply to the Court for leave to cross-examine the expert on his report, and on that application the Court shall make an order for the cross-examination of the expert by all the parties either—

(*a*) at the trial, or
(*b*) before an examiner at such time and place as may be specified in the order.

Remuneration of Court expert (O.40, r.5)

40/5   5.—(1) The remuneration of the Court expert shall be fixed by the Court and shall include a fee for his report and a proper sum for each day during which he is required to be present either in Court or before an examiner.

(2) Without prejudice to any order providing for payment of the court expert's remuneration as part of the costs of the cause or matter, the parties shall be jointly and severally liable to pay the amount fixed by the Court for his remuneration, but where the appointment of a court expert is opposed the Court may, as a condition of making the appointment, require the party applying for the appointment to give such security for the remuneration of the expert as the Court thinks fit.

Calling of expert witnesses (O.40, r.6)

6. Where a Court expert is appointed in a cause or matter, any **40/6**
party may, on giving to the other parties a reasonable time before
the trial notice of his intention to do so, call one expert witness to
give evidence on the question reported on by the Court expert but
no party may call more than one such witness without the leave of
the Court, and the Court shall not grant leave unless it considers the
circumstances of the case to be exceptional.

Scope of Order—This Order reproduced the former O.37A The object of **40/1-6/1**
the Order is presumably to enable the parties to save costs and expenses in
engaging separate experts in respect of a technical or scientific question which can
be resolved fully, quickly and comparatively cheaply by an independent expert
appointed by the Court, and also possibly to prevent the Court being left without
expert assistance in cases in which the experts of the parties may well be giving
entirely contradictory evidence on technical or scientific questions. Nevertheless,
the Court expert can only be appointed under this Order, on the application of a
party to the suit.

The definition of 'expert' in r.1(4) includes scientific persons, medical men,
engineers, accountants, actuaries, architects, surveyors and other speciallly skilled
persons. As to who is qualified to give expert evidence of foreign law, see Civil
Evidence Act 1972, s.4(1).

For an estimate of the Court expert, see *per* Lord Denning MR in *Re Saxton*
[1962] 1 WLR 968; [1962] 3 All ER 92, in which he expressed the hope (not perhaps
limited to a legally aided party) that in future careful consideration may be given
to the appointment of a Court expert.

For the inherent power of the Court to appoint its own expert, apart from this
Order, see *Kennard* v. *Aslam* (1894) 10 TLR 213; *Henson* v. *Ashby* [1896] 2 Ch1 p 26;
and see *per* Lord MacNaughton in *Coles* v. *Home and Colonial Stores Ltd* [1904] AC
179 p 192, and *Badische etc.* v. *Lewisham* (1883) 24 ChD 156.

Compare also O.33, r.6, which enables the trial Judge to sit with assessors.

Practice—Applications under this Order have been very few in number **40/1-6/2**
excepting orders by Official Referees. The following is a brief account of the
practice adopted where the parties had failed to agree upon an expert. After an
order was made which specified the questions with which the expert was to deal,
the Masters' Secretary communicated with an expert, sending him a copy of the
the Order. Upon his consenting to act, his name was inserted in the order. He then
made a report on the subject-matter without having seen the pleadings. 14 days
elapsed, the Masters' Secretary, having inquired of the expert as to his fee, called
upon the solicitors to agree the amount. On their failing to do so the summons was
restored, notice being given to the expert who was asked to attend in person or
to write to the Court. On the return to this summons the Master fixed the fee, and

indorsed it upon the original order, together with a direction to each party to pay half the amount into Court, there to remain pending the trial of the action or further order.

Table of Cases

Practice Direction:
Ollett *v.* Bristol Aerojet Ltd (Practice Note) [1979] 1 WLR 1196

Table of Statutes

Table of Rules of the Supreme Court (RSC)

EC Directives

Index

site investigation, 84
standard of skill, 49
state of the art, 8, 51, 182
statement of claim, 82
statutory obligations, 38, 43, 55
Supply of Goods and Services Act
    1982, 55
Supreme Court Rules *see* Rules of
    Supreme Court
style of report, 5

thoroughness, 61
torts, 38, 44, 56

trial, 172
    checklist, 178
    cross-examination, 175
    examination-in-chief, 174
    procedure, 173
    re-examination, 177

uneconomic working, 42
Unfair Contract Terms Act 1977,
    49, 57, 184

weaknesses identified, 18
weight of evidence, 23
without prejudice meeting, 16